FABLESS:

The Transformation of The Semiconductor Industry

DANIEL NENNI
PAUL MCLELLAN

WITH FOREWORD BY CLIFF HOU,
VP OF R&D, TSMC

A SEMIWIKI.COM PROJECT

Fabless: The Transformation of the Semiconductor Industry
Copyright 2013 by SemiWiki.com LLC. All rights reserved. Printed in the United States of America. Except as permitted under the United States Copyright Act of 1976, no part of this publication may be reproduced or distributed in any form or by any means, or stored in a data base or retrieval system without the prior written consent of the publisher.

Authors: Daniel Nenni and Paul McLellan
Editors: Beth Martin and Shushana Nenni
ISBN-13: 978-1497525047
ISBN-10: 1497525047
BISAC: Business & Economics / General

Fabless: The Transformation of the Semiconductor Industry

Table of Contents

Foreword v
Preface vii
Chapter 1: The Semiconductor Century 11
Chapter 2: The ASIC Business 23
   In Their Own Words: VLSI Technology 31
   In Their Own Words: eSilicon Corporation 38
Chapter 3: The FPGA 47
In Their Own Words: Xilinx 55
Chapter 4: Moving to the Fabless Model 67
   In Their Own Words: Chips and Technologies 71
Chapter 5: The Rise of the Foundry 75
In Their Own Words: TSMC and Open Innovation Platform 83
   In Their Own Words: GLOBALFOUNDRIES 93
Chapter 6: Electronic Design Automation 103
   In Their Own Words: Mentor Graphics 113
   In Their Own Words: Cadence Design Systems 127
   In Their Own Words: Synopsys 142
Chapter 7: Intellectual Property 157
   In Their Own Words: ARM 164
   In Their Own Words: Imagination Technologies 172
Chapter 8: What's Next for the Semiconductor Industry? 185

Fabless: The Transformation of the Semiconductor Industry

Foreword

Dr. Cliff Hou, Vice President, Research and Development, TSMC

Semiconductor innovation has the power to change the world. Although, well over half a century ago, when semiconductors first came into being, few people really saw that promise. That power of semiconductors to innovate has stretched beyond its original applications. It also has changed how semiconductors are manufactured.

Over the first 30 years of its existence, the semiconductor industry followed the proven integrated manufacturing model of the time. Those companies who owned the manufacturing assets made, marketed, researched and developed their own products. But then, the dynamics of innovation mingled with laws of supply and demand and a new concept—outsourcing—emerged and gave birth to what is known today as the dedicated foundry model, and the world has never been the same.

Dr. Morris Chang is credited with identifying the innovation need and providing the resources to meet it. The need was making available manufacturing resources that are 100 percent dedicated to those emerging semiconductor companies that lacked the financial wherewithal to own their own expensive equipment. Like all great ideas, the premise was simple. What no one foresaw is that it would give rise to two, if not three, new industry segments, all of which contribute greatly to the innovative spirit of the industry today.

When Dr. Chang established the Taiwan Semiconductor Manufacturing Company (TSMC) in 1987, the foundry segment and the fabless semiconductor model were born. Today, fabless semiconductor companies—those companies who do not own manufacturing resources—are the fountainhead of innovation that is the foundation for our electronic world. The foundry segment has allowed these companies to invest in design and innovation rather than in manufacturing. As a result, innovation and the world economy have raced forward at an unprecedented pace. This has given nearly every semiconductor company the flexibility to innovate widely and creatively, constantly expanding the universe of products we rely upon today.

Equally remarkable has been the rise of a powerful design ecosystem to complement the fabless industry. The ecosystem works in unison with designers and foundries to ensure that the IP, design tools, and services needed to get next-generation designs taped-out and in production are proven and ready to help customers meet their time-to-market goals. Today, the emergence of the fabless model, the dedicated foundry industry segment and an independent design ecosystem are driving the mobile revolution and will be the foundation of the internet-of-things.

Even as this book was being written, the semiconductor industry continued to evolve. The drive to integrate the design and manufacturing links in the semiconductor value chain is now being extended downstream (to manufacturing equipment and materials suppliers) and upstream to major product companies. This is taking on the power of integration—virtual integration. Virtual integration is, by definition, the power of collaboration that blazes the direction and vision for the next generation of innovation.

Innovation will always be the hallmark of the semiconductor industry and it is the theme that runs through this book. I'm honored and humbled to be part of this exciting industry and equally honored and humbled to offer my comments as the introduction to this book.

Dr. Cliff Hou
January 2014

Preface

The purpose of this book is to illustrate the magnificence of the fabless semiconductor ecosystem, and to give credit where credit is due.

We trace the history of the semiconductor industry from both a technical and business perspective. We argue that the development of the fabless business model was a key enabler of the growth in semiconductors since the mid-1980s. Because business models, as much as the technology, are what keep us thrilled with new gadgets year after year, we focus on the evolution of the electronics business.

We also invited key players in the industry to contribute chapters. These "In Their Own Words" chapters allow the heavyweights of the industry to tell their corporate history for themselves, focusing on the industry developments (both in technology and business models) that made them successful, and how they in turn drive the further evolution of the semiconductor industry.

Before we dive in, let's define some terms. Rather than electronics, which refers to whole devices like your cell phone or TV, we'll be using the terms chip, IC, ASIC, SoC, and FPGA throughout the book as we focus on the components that go into the devices. Chip or IC can refer more broadly to the two main types of semiconductor devices we cover: ASICs and SoCs (systems-on-chip), and FPGAs (field-programmable gate arrays). We have chosen not to cover many other electronic components including memory, flash, mixed-signal technology, and micro-electro-mechanical systems (MEMS).

We also talk about several phases of development in the semiconductor industry, and use the following terms to describe the companies and technologies that define a particular business model.

IC: An integrated circuit, also called a chip, is a set of electronic circuits, including transistors and other components, on a silicon substrate.

Systems company: A systems company makes a consumer product from chips that other companies have designed. Examples include Cisco and Apple.

Semiconductor company: Also called integrated device manufacturer (IDM), these companies, like Intel and Samsung, design and manufacture standard ICs that systems companies use in their products. Until the mid-1980s, all semiconductor companies were IDMs, that is, they controlled both the design and manufacture of their chips. This changed gradually, and now there are only a few (Intel and Samsung notably). All other chip makers outsource the manufacturing of their designs to a foundry.

ASIC: Application specific integrated circuit refers to two things: a chip that is custom designed for a specific application, rather than for a general-purpose application, and to the type of company that developed in the 1980s that performed the physical design and manufacturing of these application-specific ICs for other semiconductor or systems companies. "ASIC" is now commonly used interchangeably with "IC."

SoC: A system-on-chip is an IC that integrates all components of a computer or other electronic system into a single chip. It may contain digital, analog, mixed-signal, and often radio-frequency functions—all on a single chip substrate.

Fabless company: A company that designs their own chip but outsources the manufacturing to a third-party, either a pure-play foundry or an IDM that sells excess fab capacity. This is the prevailing business model today.

EDA: Electronic design automation companies make the software that is used to design all modern semiconductor devices. The three dominant EDA companies today are Synopsys, Cadence Design Systems, and Mentor Graphics.

IP: Semiconductor intellectual property companies sell chip designs that are implemented in their customer's ASICs, SoCs, or other semiconductor devices. A useful metaphor is that rather than selling a complete house, IP companies sell you the blueprint. The best known IP company is ARM.

Foundry: A business that is a dedicated semiconductor fabrication facility that does not design its own ICs. The term "fab" refers to any semiconductor fabrication plant, whether run as part of an IDM (like Intel) or as a foundry (like TSMC).

The economics of designing a chip and getting it manufactured is similar to how the pharmaceutical industry gets a new drug to market. Getting to the stage that a drug can be shipped to your local pharmacy is enormously expensive. But once it's done, you have something that can be manufactured for a few cents and sold for, perhaps, ten dollars. ICs are like that, although for different reasons. Getting an IC designed and manufactured is incredibly expensive, but then you have something that can be manufactured for a few dollars, and put into products that can be sold for hundreds of dollars. One way to look at it is that the first IC costs many millions of dollars—you only make a lot of money if you sell a lot of them.

What we hope you learn from this book is that even though IC-based electronics are cheap and pervasive, they are not cheap or easy to make. It takes teams of hundreds of design engineers to design an IC, and a complex ecosystem of software, components, and services to make it happen. The fabs that physically manufacture the ICs cost more to build than a nuclear power plant. Yet year after year, for 40 years, the cost per transistor has decreased in a steady and predictable curve. There are many reasons for this cost reduction, and we argue that the fabless semiconductor business model is among the most important of those reasons over the past three decades.

The next chapter is an introduction to the history of the semiconductor industry, including the invention of the basic building block of all modern digital devices, the transistor, the invention of the integrated circuit, and the businesses that developed around them.

Chapter 1: The Semiconductor Century

Although the technology behind our electronic devices is largely hidden from sight, its influence on our daily lives, our health, our economy, and our entertainment is undeniable. Today, digital electronics are ubiquitous and indispensable to the daily life of modern people. But it wasn't always so.

Two big things happened to bring consumer electronics into every household: the invention of the transistor in 1947, and the invention of the integrated circuit (IC) in 1959. Then, lots of little things happened to make ICs small and cheap enough to occupy nearly every aspect of our lives.

For the average western child in the 1950s and 1960s, the only electronics in the household were the radio and the television, both of which contained tubes (valves in some countries) not digital semiconductor technology. The only widespread electronic product was the transistor radio, which you could buy for roughly $20 ($150 in 2013 dollars).

In the 1970s, kids still watched analog TVs, but all radios were transistor based and you could buy a pocket calculator (for about $160 in 2013 dollars), an early PC, digital watches, and an Atari game console. A kid in the 1980s would also have a Walkman, a CD player, a VCR, video camera, boom box, an electric typewriter, and maybe an actual IBM PC. Anyone born after 1990 will probably not remember a time without cell phones, flat panel TVs, GameBoys, laptops, and tablets. Electronics are now incorporated into nearly everything from home thermostats and toothbrushes, to cars and medical devices.

Today, an iPad has more processing power than a Cray supercomputer in 1990, which was the size of a refrigerator. Our cars contain dozens of microprocessors. We shop online. We read books on tablets. We play video games on consoles that are more powerful than the flight simulators of twenty years ago. We'll let futurists predict what electronics a child born in 2013 might never live without. It's been a steep curve up and to the right for the number and types of electronic devices we encounter daily.

The Invention of the Transistor and the Integrated Circuit

The transistor, which is just a switch that controls the flow of electrical current in a computer chip, is at the heart of almost all electronics. This makes it among the most important inventions of the 20th century. It was invented at Bell Labs in New Jersey in 1947 by John Bardeen, Walter Brattain, and William Shockley. Shockley then left Bell Labs and returned to Palo Alto, CA, where he had been brought up. He opened Shockley Semiconductor Laboratory as a division of Beckman Instruments, and tried to lure ex-colleagues from Bell Labs to join him. When he was unsuccessful, he searched universities for the brightest young graduates to build the new company. This was truly the genesis of Silicon Valley and some of its culture that still exists today. Shockley is credited with bringing the silicon to Silicon Valley.

> *"What we didn't realize then was that the integrated circuit would reduce the cost of electronic functions by a factor of a million to one, nothing had ever done that for anything before"* - Jack Kilby

Shockley's management style was abrasive and he alienated many who worked for him. The final straw came when Shockley decided to discontinue research into silicon-based transistors. Eight people, known as the "traitorous eight," resigned and with seed money from Fairchild Camera and Instrument they created Fairchild Semiconductor Company. Almost all semiconductor companies, notably Intel, AMD, and National Semiconductor (now part of Texas Instruments) have their roots in

Fairchild in one way or another. For this reason, they were referred to as "Fairchildren." These companies drove the development of silicon-based integrated circuits. Silicon wasn't the only material in play for making transistors, but it turned out to be the winning technology.

The next key invention came in 1959 from Jean Hoerni at Fairchild when he created the "planar" manufacturing process, which flattened the transistor and allowed it to be mass-produced. The same year, Jack Kilby at Texas Instruments and Robert Noyce at Fairchild developed the integrated circuit. The IC connected diodes, transistors, resistors, and capacitors on a single silicon chip. Kilby and Noyce both received the National Medal of Science, and Kilby received the Nobel Prize for the work in 2000 (Noyce died in 1990).

The integrated circuit turned out to be the big breakthrough. Until that point, transistors were built one at a time and wired together manually using "flying-wire" connections. The planar manufacturing process allowed multiple transistors to be created simultaneously and connected together simultaneously. By 1962, Fairchild was producing integrated circuits with about a dozen transistors. Much has changed in the intervening years, but we use the same basic principle to build modern billion-transistor chips. Those two inventions, the transistor and then the integrated circuit, are the key to electronics today and all the ways in which electronics affects our lives.

Moore's Law

> "The whole point of integrated circuits is to absorb the functions of what previously were discrete electronic components, to incorporate them in a single new chip, and then to give them back for free, or at least for a lot less money than what they cost as individual parts. Thus, semiconductor technology eats everything, and people who oppose it get trampled." -Gordon Moore

In 1965, Gordon Moore was the head of research and development at Fairchild. Moore noticed that the number of transistors on the integrated circuits that Fairchild was building seemed to double every two years, as shown in the graph from Moore's original 1965 article in *Electronics* (vol.

38, number 8) titled, "Cramming More Components onto Integrated Circuits." As he pointed out there, "Integrated circuits will lead to such wonders as home computers, automatic controls for automobiles, and personal portable communications equipment."

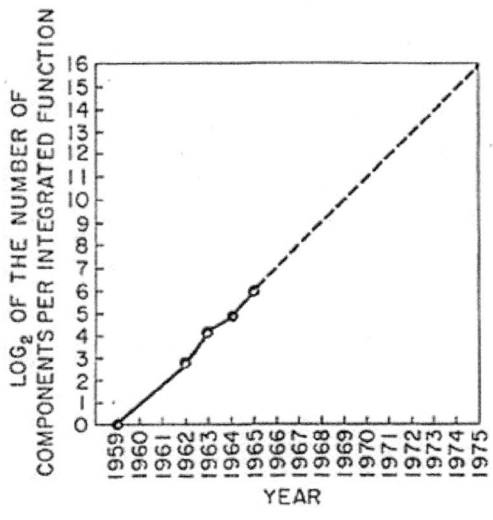

Fig. 2 Number of components per integrated function for minimum cost per component extrapolated vs time.

Moore's original graph predicting a steady rise in the number of transistors on a chip.

Remember that this was 1965, when an integrated circuit contained 64 transistors. This was an extraordinary prediction. And he was right; we do have home computers, automatic controls for automobiles (not quite fully automatic yet), and personal portable communications equipment also known as cell-phones. His prediction differed from popular science fiction assumptions about future technology because it was based on observed facts about the trajectory of computational capacity. Notice that he did not predict flying cars or unlimited power sources, two technologies that were assumed to be inevitable by mid-century futurists. Surprisingly, nearly 50 years after Moore made his observation, semiconductors seem still to be increasing in complexity at this rate. Gordon Moore's original prediction is now known as "Moore's Law."

However, it is possible to look at Moore's Law another way: the cost of any given *functionality* implemented in electronics halves every two years or so. Over a period of twenty years, this is a thousand-fold reduction. A modern video-game console has far more computing power and much better graphics than the highest-end flight simulators of the 1970s. Every ink-jet printer has far more computing power than NASA had at its disposal for getting to the moon.

It is this exponential reduction of electronic costs that has transformed so many aspects of our lives in the last twenty years or so since integrated circuits became cheap enough to go into consumer electronic products. Because of this fast growth in semiconductor technology, we have certain expectations about electronics that we don't have for anything else. We don't expect our cars to cost half as much or get double the gas-mileage every few years. Intel made another comparison: if the airline industry obeyed Moore's law, a flight from New York to Paris taking seven hours and costing $900 in 1978 would have taken a second and cost a penny in 2005.

How ICs are Made

The process of designing and manufacturing an IC can seem abstract. In fact it is complex, but not unfathomable. The design of ICs used to be a manual task, but is accomplished now with the help of specialized software. That aspect will be covered later. The basic manufacturing technique has evolved from the original planar process, in which ICs are built up in layers on top of a disc of silicon called a wafer. A modern wafer is 12" in diameter (300 mm) with an area of roughly 70,000 sq mm, about the size of a dinner plate. If the ICs are small, say 1 mm on each side, the wafer will hold 70,000 of them. If you're making giant ICs, say 20x20 mm, you can fit only 148 on a wafer. The ICs on a wafer are called die. Die is used as both singular and plural in the semiconductor industry.

Starting with the bare silicon wafer, layers of different materials—semiconductor, metal, and dielectrics—are deposited one at a time. The layers that define the transistors are created first, then all the transistors are created. Next, the layers of metal are deposited and then etched with chemicals to define them into the wires that hook all the transistors together and to supply power from outside the chip (from the battery in your cell-phone, for example).

The key feature of the production process is that all the transistors on all the die on the wafer are created simultaneously, and each layer of metal is created simultaneously across the whole wafer. It is this incredible level of efficiency, making trillions of transistors at once, that has allowed the price of electronic products to fall by around 5% per month, year after year.

The manufacturing process is based on a photographic process known as photolithography in which each die is exposed to light through a mask (more correctly called a reticle). The reticles are usually the negative image of all the components of the integrated circuit. A machine called a stepper exposes each die one at a time to a flash of light from a laser through the reticle, and then steps over to the next die until the whole wafer has been exposed. The photographic process captures the mask pattern on photoresist, a wafer coating whose chemical properties are modified by exposure to the light source through the reticle. The wafer is then developed, resulting in the corresponding reticle pattern in photoresist at each die location.

Light shines through a reticle, which acts as a stencil to create patterns on the wafer. Illustration courtesy of Intel.

The huge gain in efficiency comes after the stepper is done stamping the pattern onto each die. That's when the entire wafer is processed (etched, doped, heated, etc.) to transform the patterns into the real transistors, wires, and vias that connect the metal on different layers that make up the final integrated circuit.

It is worth emphasizing that the manufacturing process doesn't depend on what is being manufactured. A computer printer doesn't need to be reconfigured depending on what you want to print, you just send it different data. In the same way, a semiconductor manufacturing process doesn't depend on what the circuit is going to do.

The full details of the manufacturing process are obviously too complex to go into here. The important thing to remember is that it doesn't matter how many transistors are on the die, or what the final product will be—all transistors on a die are created at once and all die on the wafer are processed very efficiently at the same time.

Where ICs are Made

The factories that make ICs are called fabs. Inside the fab is kept very clean—a hospital operating theater is filthy by the standards of the "clean rooms" in a fab. The air in the fab may be completely changed every few seconds, as high-efficiency particulate air (HEPA) filters in the ceiling blow air down and out through perforations in the floor before being filtered and recirculated. In fact, recently, fabs have found that even that air is not clean enough. Even a few random particles landing on a die can ruin it. These days, the wafers being processed are contained in even cleaner boxes that attach to each piece of manufacturing equipment in turn. A large part of the cost of a fab is not the manufacturing equipment, expensive though it is, but the equipment for keeping everything inside the fab clean.

Inside an Intel fab, a technician works in a clean suit, or "bunny suit." Image courtesy of Intel.

Why is cleanliness so important? The transistors on a modern integrated circuit are 20 nanometers (nm) across. There are 1 million nanometers in a millimeter. By contrast, a human hair is around 100,000 nm. Obviously a hair ending up on a wafer would be a complete disaster, blocking thousands of transistors from being manufactured correctly and causing that die to fail. But it only takes something around 10 nm across to fall on the wafer to cause a die to (probably) fail. If a die is not manufactured correctly, it is simply thrown away. There is typically no repair process to fix it after its made.

A modern fab is wildly expensive. One major company estimated a cost of $10 *billion* dollars for the fab due to start construction in 2014. Since it has a lifetime of perhaps 5 years, owning a fab costs around $50 per second, and that's before you buy any silicon or chemicals or design any chips. Obviously, anyone owning a fab had better plan on making and selling a lot of chips if they are going to make any money. That's exactly what they do: a modern fab manufactures over 50,000 dinner-plate sized wafers every month.

Fabs were not always so expensive and until relatively recently, most semiconductor companies owned their own fabs. In 1980 there were no semiconductor companies that didn't own their own fabs to manufacture their own designs. However, the economics of fabs has completely changed the semiconductor ecosystem over the last twenty years or so. The model for semiconductor companies now is to outsource the manufacturing. Companies that do this are called 'fabless' and the companies that manufacture their ICs are called foundries. This change in the semiconductor ecosystem is a recurring theme of this book and has been essential to the success of the semiconductor industry.

Business Models from Fab to Fabless

The first step that led to the outsourcing of manufacturing was when companies began sharing their in-house fabs with other companies. A company with a large fab would have excess capacity at times. To keep the lines busy, they sold that capacity to other companies who needed more.

Then, in the early 1980s, a new type of semiconductor company formed that specialized in helping systems companies design just the right chip for their application, as opposed to buying standard ICs off the shelf. These new companies would supply the knowledge of physical chip design and also manufacture the chips (or have them manufactured) and ship them back to the systems companies. These chips were known as application specific integrated circuits or ASICs (although the less catchy term "customer specific integrated circuits" would have been more correct). The ASIC model allowed companies to design custom integrated circuits without having to maintain the infrastructure of a fab.

By the mid-1980s, more companies started making specialized ICs, but without investing in their own fabs. Instead, these companies would purchase excess foundry capacity from other fabs. These companies came to be called, for obvious reasons, fabless semiconductor companies. This is when semiconductor companies with fabs became known as IDMs (integrated device manufacturer), to distinguish them from the fabless companies.

In 1987, another new breed of semiconductor company was created: the pure-play foundry. A pure-play foundry only manufactures ICs for other companies who are either fabless or had limited capacity in their own fabs. They do not design semiconductor products themselves. Before the foundry business came along, getting a semiconductor company off the ground was difficult and expensive. Building a fab was expensive, and starting a fabless semiconductor company required a complicated negotiation for excess foundry capacity at a friendly IDM. Once foundries arrived, the cost and the risk of entering the semiconductor market lowered drastically. The result? A surge of new fabless semiconductor companies in the 1990s, many funded by Silicon Valley venture capitalists to address the growing markets for computer graphics, networking chips, and wireless phone chips.

The move to a fabless model wasn't universally hailed as a good idea. Jerry Sanders, the co-founder and long-time CEO of Advanced Micro Devices (AMD), famously noted in the late 1980s as the fabless revolution was getting underway, that, "Real men have fabs." What he meant was that design and process needed to be tightly coupled. Because AMD was competing with Intel in the microprocessor business, this statement was possibly true for his business. It turned out not to be true for many businesses.

Over time, another change happened. As the specialized knowledge about how to design integrated circuits gradually spread, many systems companies stopped using the ASIC companies in favor of doing their designs in-house. By the 1990s, many systems companies had very large integrated circuit design teams and the ASIC companies gradually started selling more and more of their own products until they became, in effect, IDMs.

As fabs got more expensive, more IDMs (like Texas Instruments and AMD) also chose the fabless model. Some switched to being completely fabless, others kept their own older fabs and used the third-party foundry for the most advanced ICs. This was known as fab-lite.

This is the landscape today. There are a few IDMs such as Intel who design almost all of their own chips and build them in their own fabs. There are foundries who design no chips, they only manufacture them for other companies. Then there are fabless semiconductor companies such as Xilinx and Qualcomm along with their fab-lite brethren such as Texas Instruments, who design their own chips, sell their own products, but use foundries for all or part of their manufacturing.

Along the way, there have been other players that helped bring semiconductor technology and business to the current state. One is called electronic design automation (EDA), which is the specialized software that's needed to design ICs. This software was once developed in house by each semiconductor company, but was later outsourced. The same is true for the components that go onto many ICs, or systems-on-chips (SoCs). Semiconductor companies once had to make all the components that went on their chips themselves, or have them custom made by another company. Now there is a robust market for licensing a wide variety of off-the-shelf functions to put on chips. These include things like A/D converters, memory, and processors, and are collectively known as silicon intellectual property, or IP.

From the IC to the iPad

With this basic history of the transistor, we can look at the changes in the semiconductor business and technology through the years and see how we've arrived at the current state of the industry. The chapters of this book cover main story arc of the semiconductor industry:

- Genesis: the invention of transistor and the integrated circuit
- The first major transition from off-the-shelf components to ASICs
- The second major transistion from owning fabs to the fabless model
- The growth of EDA: selling the software that makes it all work

- The role of IP: selling the building blocks for chips
- The future: industry luminaries look to what comes next

Each main topic is presented in a chapter that explores the history and key technologies. Each chapter is punctuated by sections that were contributed by the leading companies in the fabless semiconductor landscape today. They explain in their own words their history and role in the larger ecosystem. The last chapter of the book passages from industry luminaries who share their vision of what will take the semiconductor industry to the next level of innovation and financial success. Our hope is that this combination of objective and subjective histories is both informative and entertaining.

Chapter 2: The ASIC Business

Before the 1980s, ICs contained a limited number of transistors and were designed and created by the traditional semiconductor companies like Fairchild and Texas Instruments. The chips were generic; basic building blocks that everyone bought and made into products. However, by the early 1980s, as semiconductor technology reached a point where much more functionality could be fit onto a single chip, the people who made electronics products began to search for new ways to stand out from the competition. They wanted ICs that were differentiated from the competition, and that were tuned to work specifically in their products. This drove the development of a new type of chip, the application-specific integrated circuit or ASIC, and a new business model that drastically changed the layout of the semiconductor industry.

Traditional Semiconductor Business Stalls

The business model of semiconductor companies from the beginning of the IC until the 1980s was to imagine what the market needed, create it, manufacture it and then sell it on the open market to multiple customers. As electronic products became more sophisticated, their customers wanted chips that more specifically met their needs, rather than the generic chips that were available to everyone. This was something the traditional semiconductor companies were not equipped to provide for both business and technical reasons. On the technical end, semiconductor companies knew a lot about semiconductors, but they lacked system knowledge, and so were unable to design specific ICs for every market segment. On the business side, providing more versions of their products would increase design overhead costs and reduce the advantages of manufacturing huge volumes of a limited number of products. The systems companies, on the other hand, knew exactly what they wanted to build but didn't have enough semiconductor knowledge to create their own chips and didn't have a means to manufacture those chips even if they could design them. The systems companies needed a new way of doing chip design.

The ASIC Business Blooms

There was clearly a new niche forming for a business that could figure out how to create custom ICs for systems companies. Two companies in particular, VLSI Technology and LSI Logic, pioneered this new ASIC business. They both applied deep knowledge of semiconductor design and manufacturing to a business model that consisted largely of building other people's chips. What emerged was a model in which the systems companies did the early part of the design (called front-end) that specifies the exact functionality they want, then handed the physical design (called back-end) and manufacturing responsibilities to the ASIC company.

While it was initially thought of as a terrible business to be in—high engineering costs and few customers—the advantages to this new model became evident and the new ASIC companies did very well. LSI Logic, for example, reported revenues of $2.75 billion by 2000.

The new ASIC model set the stage for a cascade of changes to the semiconductor industry. For example, the budding electronic design automation companies took note of this new market. They realized that their design automation systems used for printed circuit boards could also be used for the front-end steps of ASIC design too.

How the ASIC Design Model Works

ASIC design typically worked like this: A systems company, typically one building an add-on board for the PC market that was the big driver of electronics in that era, would come up with some idea for a new chip.

They would negotiate with several ASIC companies and choose one to work with even though they only had a vague idea of the size of the design at that point. The chosen ASIC company would supply them with a library of basic building blocks called standard cells.

The systems company would use a software tool called a schematic editor to create the design, picking the cells they wanted from the library and deciding how those cells should be connected. The output from this process is called a netlist, essentially a list of cells and connections.

Just like writing software or writing a book, the first draft of the design would be full of errors. But with semiconductor technology, it isn't possible to build the part and see what the errors are. Even back then, manufacturing the first chip, known as the prototype, could cost tens of thousands of dollars and take a couple of months. Also unlike book writing, it's not possible to simply proofread or inspect the schematic; too many errors would still slip through. Instead, designers simulated the function of the design with software. A flight simulator tells a pilot what would happen if he or she moves the controls a certain way, and there is no cost to crashing in the simulation. In the same way, a simulation of the design checked how it behaved given certain inputs without requiring the expense of building the chip. Errors detected through simulation could be fixed and the simulation could be run again until no more errors remained.

When the design was finally determined to be functionally correct, the netlist was sent from the systems company to the ASIC company for the next step. Using a software program that placed the standard cells and wired them together (known as place & route), the netlist would be converted to a physical layout. The netlist is a list of cells and connections, something like an architectural spec that says which room connects to which and by how many doors; the output of place & route adds the physical locations and specific wire routing, analogous to a completed house blueprint. In addition to creating the actual layout that will be manufactured, this process also created a detailed account of timing—how long every signal takes as it travels from its source to the transistor it switches on or off. This detailed timing was sent back to the systems company for a final simulation to ensure that everything still worked.

After the design passed final simulation, the systems company took a deep breath and gave the go-ahead to manufacture prototypes of the chip. At the time, all the design data needed to make the photomask was written onto a computer tape, so the process was, and still is, called tape-out.

The ASIC company then had the masks made that were needed to run the design through their fab. Chips were manufactured using one of two main ASIC technologies; gate-array or cell-based. In a gate-array design, the gates—a group of transistors that perform a function—were pre-fabricated on a wafer (the gate-array "base") so the masks only pattern the interconnect. In cell-based design, masks were required to pattern all layers on a blank wafer. The gate-array approach was faster and cheaper, but less flexible. It was faster, because there were fewer masks to make and fewer layers to be manufactured. It was cheaper, because the gate-array bases were mass produced in higher volume than any individual design would be. However, gate-array substrates only came in certain fixed sizes, and so the designs often left many potential gates unused.

In a couple of months, the prototypes would be finished and samples shipped back to the systems company. These parts would then be incorporated into complete systems and those systems tested. For example, if the chip went into an add-in board for a PC, a few boards would then be manufactured, put into a PC, and checked for correct operation.

At the end of that process, the systems company took another deep breath and placed an order with the ASIC company for volume manufacturing; requesting thousands, or possibly even millions, of chips. They would receive these a few months later, build them into their own products, and ship those products to market. The final step in this journey was the day we, the consumers, brought home our very own personal computer or CD player.

The Lasting Effect of the ASIC Model

All semiconductor companies were caught up in ASIC in some way or another because of the basic economics. Semiconductor technology allowed systems companies to make medium-sized designs, and medium-sized designs were pretty much all different. The technology didn't yet allow whole systems to be put on a single chip. This meant that semiconductor companies could no longer survive by just supplying basic building-block chips because those were largely being superseded by ASIC chips. But they also couldn't build whole systems like a PC, a television, or a CD player because semiconductor technology did not allow for that level of integration. Eventually, most semiconductor companies, including Panasonic, Fujitsu, and Intel, joined the ASIC business, thus making the market very competitive.

Although the ASIC business model filled an important niche in the development of electronic products, it turned out to be a difficult business in which to make money. The systems company owned the specialized knowledge of what was in the chip, so the semiconductor company could not price to value. The systems company also knew the size of the chip and thus roughly what it should cost to make. The best money for ASIC companies turned out to be making the largest, most difficult designs. It took more expertise to successfully complete the physical design of these big designs, so the leading ASIC companies, VLSI Technology and LSI Logic, could charge premium pricing based on their ability to complete the most challenging designs on schedule. If you are building a sky-scraper you don't go with a company that has only built houses.

ASIC companies had few designs they could make money on, and it gradually became obvious just how unprofitable low-volume designs were. All the ASIC companies realized that there were less than a hundred designs a year that were really worth winning, and competition to win those became fierce.

During this time, semiconductor technology continued to advance and it became possible to build whole systems (or large parts of them) on a single integrated circuit. These were known as systems-on-chip, or SoCs. The ASIC companies started to build and sell whole systems, such as chipsets for PCs or cell-phones much like the traditional semiconductor model, alongside their traditional ASIC business. This made all semiconductor companies start to look the same, with lines of standard products and, often, an ASIC product line too.

One important aspect of the ASIC model was that the "tooling," the industry word for the photomasks, belonged to the ASIC company. This meant that any given design could only be manufactured by its specific ASIC company. Even if another semiconductor company offered them a great deal to manufacture a completed design, the systems company couldn't just hand over the masks made by a previous ASIC company. This became very important in the next phase of what ASIC would morph into: design services.

ASIC design required a network of design centers all over the world staffed with some of the best designers available, obviously an expensive proposition. Their customers started to resent paying the premium to support this infrastructure, especially on very high volume designs. While the systems companies could shop around for a better price, switching vendors was costly because it meant starting the design all over again with the new semiconductor supplier.

Eventually both VLSI Technology and LSI Logic would be acquired. VLSI was bought by NXP (still then called Philips Semiconductors) in 1999 for close to $1 billion. LSI Logic, which left the ASIC business and was renamed LSI Corporation, was acquired by Avago in late 2013 for $6.6 billion.

The ASIC Model Morphs into Design Services

By the early 1990s, in addition to the high cost of the ASIC model, two other things had changed that spelled the beginning of the end for the ASIC business. One was that foundries such as TSMC had come into existence. The second is that the knowledge of how to do physical design became more widespread and at least partially encapsulated in software tools available from the EDA industry. These changes gave systems companies a new route to silicon that bypassed the ASIC companies completely. Systems companies could now feasibly complete the entire design themselves, including the physical design, and then use a foundry like TSMC to manufacture it. This was known as customer-owned-tooling or COT, because the systems company, not the ASIC company or the foundry, owned the whole design from concept to masks. If one foundry gave poor pricing, the systems company could transfer the design to a different manufacturer without having to completely redesign the chip.

However, the COT approach was not without its challenges. Doing physical design of a chip is not a simple task. Many systems companies underestimated the value of the premium charged by ASIC companies for their expertise, and they struggled to complete designs on their own without that support. As a result, a new breed of companies, known as design services companies, emerged to meet this exploding demand for support.

Design service companies played a similar role as the ASIC companies; providing specialized semiconductor design knowledge to the systems companies. In some cases, they would do the entire design, which is called turn-key design. More often, they would do all or some of the physical design and sometimes manage the interface with the foundry to oversee the manufacturing process, another area where systems companies lacked experience. One company in particular, Global Unichip, operates with a business model identical to the old ASIC companies except in one respect—it has no fab. It uses a foundry, primarily TSMC, to build all of their customers' products.

This is the layout of the ASIC landscape today: there is very limited ASIC business conducted by a few semiconductor companies. There are design services companies and virtual ASIC companies like Global Unichip and eSilicon. There are no pure-play ASIC companies. A lot of IC functions that were once implemented as ASIC are now mostly done as field-programmable gate arrays, or FPGA, which is important enough to need a chapter of its own. The next main chapter, in fact, is an exploration of FPGAs. But first, a brief history of one of the companies that created the ASIC business model, VLSI Technology, and one of the new breed of design services companies, eSilicon.

In Their Own Words: VLSI Technology

> As one of the companies that founded the ASIC business model, VLSI Technology helped set the course of the entire semiconductor industry. The company is no longer in business, but one of their early and long-time employees, and co-author of this book Paul McLellan, has written this history of VLSI Technology.

VLSI Technology was founded in 1979 by Dan Floyd, Jack Baletto and Gunnar Wetlesen, who had worked together at the semiconductor company Signetics. The initial investments in VLSI Technology were from Hambrecht and Quist, a cross between a VC and a bank, and by Evans and Sutherland, the simulation/graphics company. Semiconductor technology had reached a point that significant systems or parts of systems could be manufactured, and the original business plan was to build a fab to manufacture parts that other people would design.

The fourth person to join the company, in 1980, was Doug Fairbairn. He was working at Xerox Palo Alto Research Center (PARC) and had started a fledgling publication on very-large scale integration (VLSI) design, Lambda Magazine. He went to interview the three founders for an article, but was intrigued by the new company.

 He immediately realized that their plans for a foundry wouldn't really work without a new generation of design automation tools.

Existing design tools of that era were polygon-based layout editors, but semiconductor technology was already past the point where you could reasonably design everything by hand. Doug decided to take the opportunity to move from the research environment into industry and create the first development group creating the next generation of software for integrated circuit design.

In the first few years of VLSI Technology, the company was sustained by designing ROMs (read-only memories) for the first generation of video game consoles, which were all cartridge-based. Each cartridge actually contained a ROM with the video-game binary programmed into it. The fab in San Jose was not yet in high volume manufacturing, and so these were actually outsourced to Rohm in Osaka, Japan. In parallel, Fairbairn hired a group of PhDs, many from Carver Mead's Silicon Structures Project at CalTech, and the profits from video games were invested in a suite of tools for what we would now call ASIC design, although that name didn't come until later.

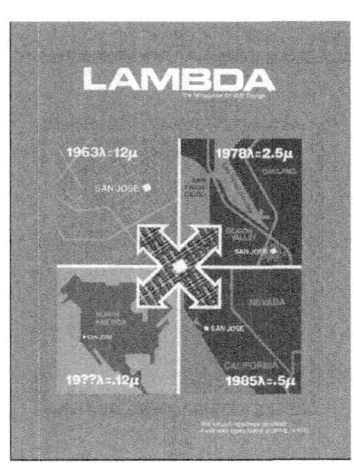

Lambda Magazine.

VLSI Technology (hereafter simply VLSI) had a fab on McKay Drive in San Jose. At the time, it was the only high-tech building in the area, surrounded by greenhouses growing flowers and, across the street, the Chrysanthemum Growers Association Hall that was sometimes used for company-wide meetings. The first process brought up was 3 μm HMOS, followed by 2 μm CMOS and 1.5 μm CMOS.

Fairly early on, the investors decided that the company's management team was too inexperienced to manage the anticipated growth. Al Stein was brought in as CEO. The company went public in February 1983, still not yet profitable, and almost immediately afterwards the three original founders departed.

The initial design technology, still based largely on the Caltech/PARC ideas in Mead and Conway's seminal book, *Introduction to VLSI Design*, was a mixture of manual design with generators for basic structures, such as registers and adders, using an internal language called VIP. The focus of the tools was on verification, with a design rule checker (DRC), a circuit extractor, a layout-versus-schematic (LVS) checker, called netcompare, and simulators—VSIM, with no timing and then TSIM, which had timing-based on a simple capacitive model.

However, designs were getting too large for this approach and despite the inelegance compared to Mead and Conway's ideas, it was clear that layout tasks had to become much more automated. This fact led them to develop standard cell libraries and a full place and route system to complement their existing schematic capture software.

In order to be successful, the design work had to get closer to the customer. Initially this meant that the customer came to VLSI, and there were several teams of customer's designers working on site at VLSI's San Jose buildings. For example, the main chip in France Telecom's initial implementation of the online service, Minitel, was created by Telic (now buried somewhere in Alcatel-Lucent) who sent a team of engineers from Strasburg, Germany to San Jose who took up residence for several months.

The next step was to create a network of design centers initially in the US, and then also in Japan and Europe, since it was clearly not scalable to bring all the customers on-site to California.

VLSI also opened a research and development site at Sophia Antipolis in the south of France. They started doing design tool development and library development, and also served as a hub of expertise to support the growing European business.

The IBM PC was then in its high growth phase and many customers of VLSI were designing products for that market (modems, add-in peripheral cards) or designing chips to create PC clones. In fact, VLSI had dozens of customers making products for the PC market. To serve these customers,

VLSI developed the first of what today is called semiconductor IP, although VLSI called them megacells (and later functional system blocks or FSBs). These included all the standard components in a PC such as the UARTs or the 6845 graphics controller.

Two key design automation products that VLSI pioneered in the late 1980s were the datapath compiler and the state-machine compiler, which was effectively one of the first synthesis tools. The datapath compiler could take a complex description for a datapath and quickly generate a fully laid out datapath on silicon, using its own optimized custom library, not standard cells. And the state-machine compiler could take a description of a state-machine (or just any old logic) and produce an optimized implementation in standard cells. Together these two tools made creating complex designs much easier.

VLSI saw robust growth in the 1980s, but it never made enough cash to fund all the investment required for process technology development and capital investment for a next-generation fab. They also had several false starts. They entered and then exited the SRAM (static memory) business. They entered and then exited a partnership to build a fab in Malaysia. They had a partnership with Philips Semiconductors licensing process technology that was never used.

In the late 1980s, they entered into a strategic partnership with Hitachi in which Hitachi gained access to VLSI's design tools and Hitachi licensed VLSI its 1 μm process technology and made significant cash investment. This meant that VLSI could bring up a competitive 1 μm technology at its second fab in San Antonio, TX. Eventually the two fabs were upgraded from 5" to 8" wafers.

Development of the Chip Set

VLSI had already developed several megacells as IP for use in PCs. A group of five engineers conducted an experiment with these megacells over a weekend that involved putting all of them together onto a few chips. This was the first PC chipset, which could be used to create a full PC with only the addition of the Intel microprocessor and memory. VLSI ran with the idea and built up a large business in PC chipset standard products to go with its mainline ASIC business.

The PC chipset business was very successful and was dominated by VLSI in the early 1990s. One generation of chipsets was even resold by Intel. However it was clear that it would eventually become a low-margin business due to competition from Asia, and probably would finally be owned by Intel who could design more and more functionality to work intimately with its own next-generation microprocessors. VLSI decided to invest in system knowledge for the GSM cellular standard that was starting to get off the ground, as well as some other attractive end markets such as digital video.

Also, in that era, around 1987, Apple decided to build the Newton personal digital assistant. They selected Acorn's RISC processor and insisted it be spun out as a separate company. So, ARM was created with Apple, Olivetti (that by then owned Acorn) and VLSI as the owners. VLSI supplied all the design tools used to design the processors, and also manufactured the initial parts. That story is told in more detail later in the book.

Meanwhile, the market for second generation (digital) GSM phones exploded. European companies, especially Nokia and Ericsson, were the most successful handset manufacturers. At one point Ericsson was accounted for 40% of VLSI's entire business. VLSI also started a major investment at its French site to develop its own GSM baseband chips. They built this up into a chipset business selling to second tier manufacturers who didn't have enough system knowledge to develop GSM baseband chips internally. They later licensed CDMA wireless technology from Qualcomm and started to develop a CDMA product line primarily for the US market. Between the standard product business and the large volume of ASIC business, especially with Ericsson, the communication segment was over half of VLSI's semiconductor business.

By 1991, it was clear that VLSI was really two companies that should already have separated: an EDA company with some of the best VLSI design tools on the market, and an ASIC/ASSP company with a network of design centers and two fabs manufacturing silicon. In 1991, the design tool business was spun out to a new company called Compass Design Automation, leaving VLSI Technology as a pure semiconductor business (with Compass as one of their EDA suppliers).

Compass struggled to shake off the perception that it wasn't really independent of VLSI and as a result it had only a small ecosystem of semiconductor companies that fully supported it with ASIC libraries. But Compass also had its own portfolio of libraries, originally developed for VLSI's ASIC business. By creating standardized design rules (called Passport) that worked in almost all fabs, it created the first library business with a portfolio of standard cells, memory compilers, the datapath compiler, and other foundation IP. This was very successful and grew to be about 30% of Compass's business.

Compass increased to nearly $60 million in revenue but it was never profitable. They had a fully integrated suite of design tools in an era when the large EDA companies, which had grown through acquisition, had educated the market to pick best-in-class point tools and use their internal CAD departments to integrate them. So Compass was swimming against the tide and despite the fact that every ASIC and every standard product made by VSLI was designed exclusively using Compass tools and libraries, they never shook off the perception that they were not leading edge. CAD groups were reluctant to standardize on Compass, at least partially because they would have much less design tool integrating to do.

In 1997, Compass was sold for $44 million to Avant!, which was mostly interested in the library business to complement their own software business. Of course, Avant! in turn was acquired by Synopsys in 2001 (for $830 million). The software part of the business, as opposed to the library development, by then was largely based in France and the entire group in France was hired by Cadence en masse where many of the individual engineers still work today. The library business was largely in California and was integrated into Avant!

VLSI's semiconductor business, both the ASIC business and the ASSP business grew through the 1990s to about $600 million in revenue. There was a focus on wireless, digital video, PC graphics and an ASIC business that was diversified into many separate segments.

In 1999, Philips Semiconductors (now called NXP) made a hostile takeover bid for VLSI Technology. Philips had struggled to bring processes to market quickly along with the required libraries. As the ASIC business got more and more consumer oriented, this became a big problem because of the very short product life-cycles. VLSI's lifeblood was ASIC and they were much quicker at getting designs going in new process generations, so Philips figured that acquiring VLSI would shake up their internal processes and also give them a network of leading design centers (by then renamed technology centers). After some negotiation, VLSI was acquired by Philips Semiconductors for just under a billion dollars and it ceased to be an independent company.

In Their Own Words: eSilicon Corporation

> *eSilicon was one of the first companies to focus on making the benefits of the fabless semiconductor movement available to a broader range of customers and markets. The company is credited with the creation of the fabless ASIC model. In this section, eSilicon shares some of its history and provides its view of the ever-changing fabless business model.*

eSilicon Corporation was founded in 2000 with Jack Harding as the founding CEO and Seth Neiman of Crosspoint Venture Partners as the first venture investor and outside board member. They both remain involved in the company today, with Harding continuing as CEO and Neiman now serving as Chairman of the Board.

Both Harding and Neiman brought important and complementary skills to eSilicon that helped the company maneuver through some very challenging times. Prior to eSilicon, Harding was President and CEO of Cadence Design Systems, at the time the largest EDA supplier in the industry. He assumed the leadership role at Cadence after its acquisition of Cooper and Chyan Technology (CCT), where Harding was CEO. Prior to CCT, Harding served as Executive Vice President of Zycad Corporation, a specialty EDA hardware supplier. He began his career at IBM.

Seth Neiman is Co-Managing Partner at Crosspoint Venture Partners, where he has been an active investor since 1994. Neiman's investments include Brocade, Foundry, Juniper and Avanex among many others.

Prior to joining Crosspoint, Neiman was an engineering and strategic product executive at a number of successful startups including Dahlgren Control Systems, Coactive Computing, and the TOPS division of Sun Microsystems. Neiman was the lead investor in eSilicon and incubated the company with Jack at the dawn of the Pleistocene epoch.

The Early Years

eSilicon's original vision was to develop an online environment where members of the globally disaggregated fabless semiconductor supply chain could collaborate with end customers looking to re-aggregate their services. The idea was straight-forward—bring semiconductor suppliers and consumers together and use the global reach of the Internet to facilitate a marketplace where consumers could configure a supply chain online. The resultant offering would simplify access to complex technology and reduce the risk associated with complex design decisions.

Many fabless enterprises had struggled with these issues, taking weeks to months to develop a complete plan for the implementation of a new custom chip. Chip die size and cost estimates were difficult to develop, technology choices were varied and somewhat confusing, and contractual commitments from supply chain members took many iterations and often required a team of lawyers to complete.

The original vision was simple, elegant and sorely needed. However, it proved to be anything but simple to implement. In the very early days of the company's existence, two things happened that caused a shift in strategy. First, a close look at the technical solutions required to create a truly automated marketplace yielded significant challenges. Soon after formation of the company, eSilicon hired a group of very talented individuals who did their original research and development work at Bell Labs. This team had broad knowledge of all aspects of semiconductor design. It was this team's detailed analysis that lead to a better understanding of the challenges that were ahead.

Second, a worldwide collapse of the Internet economy occurred soon after the company was founded. The "bursting" of the Internet bubble created substantial chaos for many companies. For eSilicon, it meant that a reliable way to monetize its vision would be challenging, even if the company could solve the substantial technical issues it faced. As a result, most of the original vision was put on the shelf. The complete realization of the "e" in eSilicon would have to wait for another day. All was not lost in the transition, however. Business process automation and worldwide supply chain relationships did foster the development of a unique information backbone that the company leverages even today. More on that later.

The Fabless ASIC Model

Mounting technical challenges and an economic collapse of the target market have killed many companies. Things didn't turn out that way at eSilicon. Thanks to a very strong early team, visionary leadership and a little luck, the company was able to redirect its efforts into a new, mainstream business model. It was clear from the beginning that re-aggregating the worldwide semiconductor supply chain was going to require a broad range of skills. Certainly, design skills would be needed. But back-end manufacturing knowledge was also going to be critical. Everything from package design, test program development, early prototype validation, volume manufacturing ramp, yield optimization, life testing, and failure analysis would be needed to deliver a complete solution. Relationships with all the supply chain members would be required and that took a special kind of person with a special kind of network.

eSilicon assembled all these skill sets. That deep domain expertise and broad supply chain network allowed the company to pioneer the fabless ASIC model. The concept was simple—provide the complete, design-to-manufacturing services provided by the current conventional ASIC suppliers, such as LSI Logic, but do it by leveraging a global and outsourced supply chain. Customers would no longer be limited to the fab that their ASIC supplier owned, or their cell libraries and design methodology.

Instead, a supply chain could be configured that optimally served the customer's needs. And eSilicon's design and manufacturing skills and supply chain network would deliver the final chip. The volume purchasing leverage that eSilicon would build, coupled with the significant learning eSilicon would achieve by addressing advanced design and manufacturing problems on a daily basis would create a best-in-class experience for eSilicon's customers.

eSilicon's positioning, DAC 2000

As the company launched in the fall of 2000, the fabless ASIC segment of the semiconductor market was born. Gartner/Dataquest began coverage of this new and growing business segment. Many new fabless ASIC companies followed. Antara.net was eSilicon's first customer. The company produced a custom chip that would generate real-world network traffic to allow stress-testing of ebusiness sites before they went live. Technology nodes were in the 180 nm to 130 nm range and between eSilicon's launch in 2000 and 2004, 37 designs were taped out and over 14 million chips were shipped.

Fabless ASIC was an adequate description for the business model as everyone knew what an ASIC was, but the description fell short. A managed outsourced model could be applied to many chip projects, both standard and custom. As a result, eSilicon coined the term Vertical Service Provider (VSP), and that term was used during the company's initial public exposure at the Design Automation Conference (DAC) in 2000.

The model worked. eSilicon achieved a fair amount of notoriety in the early days as the supplier of the system chip that powered the original iPod for Apple. The company also provided silicon for 2Wire, a company that delivered residential Internet gateways and associated services for providers such as AT&T. But it wasn't only the delivery of "rock star" silicon that set the company apart; some of the original ebusiness vision of eSilicon did survive.

The company launched a work-in-process (WIP) management and logistics tracking system dubbed eSilicon Access® during its first few years. The company received a total of four patents for this technology between 2004 and 2010. eSilicon Access, for the first time, put the worldwide supply chain on the desktop of all eSilicon's customers. Using this system, any customer could determine the status of its orders in the manufacturing process and receive alerts when the status changed. eSilicon uses this same technology to automate its internal business operations today.

Growing the Business

During the next phase of growth for the company, from 2005 to 2009, an additional 135 designs were taped out and an additional 30 million chips were shipped. Technology nodes now ranged mainly from 90 nm down to 40 nm. It was during this time that the company began expanding beyond US operations. Through the acquisition of Sycon Design, Inc., the company established a design center in Bucharest, Romania. A production operations center was also opened shortly thereafter in Shanghai, China.

Recognizing the growing popularity of outsourcing, eSilicon expanded the VSP model to include semiconductor manufacturing services (SMS). SMS allowed fabless chip and OEM companies to transition the management of existing chip production or the ramp-up and management of new chip production to eSilicon. The traditional design handoff of the ASIC model was now expanded to support manufacturing handoff. The benefits of SMS included a reduction in overhead for the customer as well as the ability to focus more resources on advanced product development.

Extensions such as SMS caused the Vertical Service Provider model to expand, creating the Value Chain Producer (VCP) model. The Global Semiconductor Alliance (GSA) recognized the significance of this new model and elected Jack Harding to their Board to represent the VCP segment of the fabless industry.

In the years that followed, up to the present day, eSilicon has grown substantially. The number of tape-outs the company has achieved is now approaching 300 and the number of chips shipped is on its way to 200 million. The company has also expanded into the semiconductor IP space. While its worldwide relationships for third-party semiconductor IP are critical to eSilicon's success, the company recognized that the ability to deliver specific, targeted forms of differentiating IP could significantly improve the customer experience.

eSilicon's first logo. The squares symbolize the end product—the chip.

Since so many of today's advanced chip designs contain substantial amounts of on-board memory, this is the area that was chosen for eSilicon's initial IP focus. The company acquired Silicon Design Solutions, a custom memory IP provider with operations in Ho Chi Minh City and Da Nang, Vietnam. This acquisition added 150 engineers to focus on custom memory solutions for eSilicon's customers.

As of June 30, 2013, eSilicon employs over 420 full-time people worldwide, of which over 350 are dedicated to engineering. Headquartered in San Jose, California, the company maintains operations in New Providence, New Jersey and Allentown, Pennsylvania; Shanghai, China; Seoul, South Korea; Bucharest, Romania; Singapore and Ho Chi Minh City and Da Nang, Vietnam. The company's diverse global customer base consists of fabless semiconductor companies, integrated device manufacturers, original equipment manufacturers and wafer foundries. eSilicon sells through both an internal sales force and a network of representatives.

The Evolving Model

The eSilicon business model has evolved further. VSP and VCP are now SDMS (semiconductor design and manufacturing services). Arguably the longest, but perhaps the most intuitive name. Through the years, eSilicon has allowed a broad range of companies to reap the benefits of the fabless semiconductor model, many of which couldn't have done it on their own.

eSilicon's current logo. The three "S" graphic symbolized the process and culture—speed, simplicity, and self-confidence.

This ability to bring a worldwide supply chain within reach to smaller companies gave eSilicon its start, but the model has worked well for eSilicon beyond these boundaries. Today, eSilicon serves customers that are much larger than eSilicon itself; customers that could "do what eSilicon does." In the early days, the company discounted its chances of winning business at an enterprise big enough to maintain an "eSilicon inside."

Time has proven this early thinking to be too limiting. Many of eSilicon's customers today can clearly maintain an "eSilicon inside," but they still rely on eSilicon to deliver their chips. Why? In two words, opportunity cost. It has been proven over time that for any enterprise the winning strategy is to focus on the organization's core competence and invest in that. All other functions should be outsourced in the most reliable and cost-effective manner possible. Simply put, eSilicon's core competency fits in the outsourcing sweet spot for many, many organizations. This trend has created new value in the fabless semiconductor sector and facilitated many new design starts.

What's Next?

As the fabless model grows, there are new horizons emerging. During its early days, the vision of using the Internet to facilitate fabless technology access and reduce risk was largely put on the shelf. The reasons included the challenges of solving complex design and manufacturing problems and the lack of a clear delivery mechanism over the Web.

Today, these parameters are changing. The Internet is now an accepted delivery vehicle for a wide array of complex business-to-business solutions. eSilicon's talented engineering team has also developed a substantial cloud-enabled environment that is used to automate its internal design and manufacturing operations every day. This team consists of many of the same people who highlighted the challenges of addressing these issues in the company's early years. What a difference a decade can make.

What if that automated environment could be made available to end users in a simple, intuitive way? New work at eSilicon is taking the company in this direction. The recent announcement of an easy-to-use multi-project wafer quote system is an example. What once could take two weeks or more, consisting of many inquiries and legal agreement reviews, is now done in as little as five minutes with an extension to eSilicon Access. With availability on both the customer's desktop and smartphone, this is clearly the beginning of a new path. eSilicon changed the landscape of fabless semiconductor in 2000 with the introduction of the fabless ASIC model. It's time to do it again and bring back the "e" in eSilicon.

Fabless: The Transformation of the Semiconductor Industry

Chapter 3: The FPGA

In the 1970s, a new type of electronic component emerged—the programmable logic device (PLD). Electronic systems until that time were generally built up out of integrated circuits called transistor-transistor logic (TTL) devices that were made by semiconductor companies like Motorola, Texas Instruments, and IBM. The TTL integrated circuits were small chips, with a handful of basic logic operations and 16-20 pins connecting it to the outside world. To make a system, like a computer or a calculator, you would attach dozens or hundreds of these small integrated circuits on to a board, perhaps along with some memory chips and a small microprocessor or microcontroller.

These early TTL logic chips were made with bipolar junction transistors, but were gradually replaced by a new technology based on metal-oxide construction (MOS). This MOS technology developed into CMOS technology (complementary metal-oxide semiconductor). CMOS, with its improved performance and lower power use, allowed for the development of the ICs we are familiar with today. CMOS became the standard technology for ASICs, and made it possible to design larger microprocessors and other large standard parts, such as display controllers and UARTs (the serial interface of the day). Unlike TTLs, ASICs also integrated all the 'glue' logic, the small bits of logic that tie the standard components together.

CMOS-based ASICs were great for high volume products, like personal computers, because the economics of semiconductors dictates that higher volume means lower per-unit cost. You need to sell a lot of ASICs to cover the high fixed costs of design and manufacturing. However, they are not as good for applications that required small numbers of parts or where the IC is extremely simple.

For these cases, semiconductor companies turned to programmable logic devices (PLDs) which, unlike logic gates that have fixed functions, are essentially blank slates that can be programmed to perform any number of tasks. With PLD technology, a semiconductor company could realize the economy of scale by manufacturing high volumes of ICs that the systems company could then customize to fit any number of different products. There were many variations on this technology, including programmable array logic (PAL), field-programmable logic array (FPLA), generic array logic (GAL), and the complex programmable logic device. In this chapter, though, we focus on the FPGAs because they have been the most successful and influential of the programmable devices.

PLDs Become FPGAs

The initial PLDs, which were brought to market in the early 1970s by Motorola, Texas Instruments, General Electric, and National Semiconductor, were very limited. One key limitation was the absence of flip-flops, circuits that have two stable states and are used to store state information. The early PLDs contained a small programmable memory that could be used to configure the device. The memory was either a PROM (programmable read-only memory), which could be programmed just once, or an EPROM (erasable programmable read-only memory), which could be programmed multiple times by erasing the old programming with ultraviolet light (the package had a small quartz window for this purpose).

While PLDs continued to develop throughout the 1970s, another piece of technology turned out to be even more important than the PLDs—gate-arrays. Gate arrays, which we covered briefly in the ASIC chapter, is an approach that uses wafers preprinted in volume with the transistors; only the wires are added later to make it function as desired. Manufacturing the gates, often called front-end-of-line or FEOL, was the slowest stage. Putting the metal connectivity on, called back-end-of-line or BEOL, was quicker.

Programmable circuits took a leap forward when the company Xilinx, founded in 1984, realized that they could combine the PLD and the gate-array approach into what became known as field-programmable gate-arrays (FPGAs). These chips contained uncommitted logic that was turned into the required functionality by programming a memory. This allows the device to be quickly programmed based on an application's requirements, and dramatically reduces time-to-market.

FPGAs used CMOS technology, which was an improvement over the original bipolar transistor technology used for the original PLDs and helped make FPGAs competitive with ASICs in low-volume applications. While FPGAs were still slower, more expensive, and consumed more power than ASICs, they didn't have the high fixed costs and time-to-market disadvantages of ASIC.

An important enabler for FPGAs was the design automation tools used to program them, which gave the FPGA programmers the same experience that ASIC designers had, initially based on schematic diagrams of the desired functionality and later using register transfer languages such as Verilog or VHDL. The complicated task of working out which memory bits to set in the array in order to get the desired functionality was completely automated.

As with PLDs, FPGAs were manufactured as identical parts in high volume and it was the system manufacturer who would configure them. Usually the memory (the PROM or EPROM) was actually a separate chip and the data would be transferred from the memory into the FPGA itself to configure it each time the system was first powered on.

How FPGAs Fueled the Fabless Business Model

The first company to focus solely on FPGAs was Xilinx, which is still the market leader today. Xilinx was also a pioneer in the fabless semiconductor business model. Instead of building a fab, as most semiconductor companies had up to that point, Xilinx leveraged personal relationships with the semiconductor division of Japan-based Seiko Epson. The first Xilinx FPGA was created in 1985 in a mature 1.2 µm manufacturing process. It ran at 18 MHz and had a 1000-gate equivalence, meaning that it could be used to implement the functionality of about 1000 gates, even though it actually contained more like 20,000 gates.

Because Xilinx didn't have its own fab, they had their FPGA manufactured by other semiconductor companies. They had contracts with a number of manufacturers to reduce the risk that they might lose supply and also to introduce competition into price negotiations. Then, in 1995, two major pure-play foundries, UMC and TSMC, opened for business. Xilinx moved all their new production to UMC, which was the start of a long relationship between the two companies.

It turns out that FPGAs were not only good business for the foundries, they were also important for helping the foundries ramp up new manufacturing processes. Early in the process development, the foundry needs designs with very high degrees of regularity; this lets them apply statistical approaches to yield improvement. Memories once served that role, but these days FPGAs are used. If an FPGA has 10,000 identical structures on it, then it is relatively easy to find systematic failures in the manufacturing process. Compare that with ASICs, which from a process point of view are pretty random.

Xilinx and UMC pioneered the "virtual IDM" relationship, where the fabless company has full access to the process technology and is an active development partner. Xilinx and UMC worked together to develop the process technology, create test chips, and so on. In fact, Xilinx had a whole floor of one of UMC's buildings for their own employees. FPGAs would usually be the first volume parts in a new technology.

The long standing relationship between Xilinx and UMC ended in 2010, when Xilinx switched their allegiance to rival foundry TSMC for the 28 nm process node. Rumors had the relationship ending as a result of 65 nm yield problems and delays in 40 nm which allowed Xilinx's main competitor, Altera, to gain significant market share. As a result of moving production to TSMC, Xilinx beat Altera to silicon on the 28 nm and 20 nm nodes which encouraged Altera to move production to Intel at 14 nm.

Use of FPGAs

Early FPGAs were used largely as "glue" logic—the gates in a design that are outside of the larger chips like microprocessors, memories and the like. As FPGAs developed, they began to be used in designs that were subject to frequent change, for example, in the networking world where standards were changing rapidly. In these cases, you don't want to wait until the standards are finalized before starting the chip design. Alternatively, you could implement the standard in software. However, because hardware still offers higher performance than software (throughput in a router, for example), FPGAs are a better choice.

For example, Cisco Systems, a company that designs and sells networking equipment, typically uses FPGAs for all but its highest performance routers. As FPGAs continued to develop, and got larger, they reached the size that entire systems or sub-systems could fit on a single chip. This meant that processors were needed on the FPGA. There are two ways this has been done: by putting an ARM or PowerPC processor (or more than one) on the FPGA as a hard macro, and by adding processors designed by the FPGA companies (for example, Altera's Nios processor) that are implemented using the FPGA fabric itself.

Today, for low-volume systems, FPGAs are the implementation medium of choice. Their big weakness, besides cost at high-volume, is that they consume a lot of power. Consequently it is not possible to use them for many mobile devices because it would make the battery life unacceptably short. But for 'tethered' systems, the flexibility and low up-front cost are very attractive.

Another use for high-end FPGAs, which are too expensive to go into any sort of consumer product, is prototyping for systems-on-chip. FPGAs are used extensively in emulation and hardware-accelerated simulation products, where the hardware needs to change to match the target design. They are well-suited to this application because verification of an SoC using only software simulation has become too time-consuming. Creating an early prototype version of the system in an FPGA, and then using that for verification, is considerably faster. While an FPGA is much slower than the actual silicon will be, it is much faster than software. Consequently, it is possible to run software loads and boot operating systems on an FPGA. As more and more systems involve complex hardware interacting with complex software, the FPGA prototyping route will only become more attractive.

FPGAs Need Design Automation and IP

Unlike with ASICs, where most of the electronic design automation (EDA) tools are provided by the 3rd party EDA industry, the software needed to design FPGAs has largely been provided by the FPGA suppliers themselves. There are several reasons for this, not the least being that the physical design tools are specific to the architecture of the fabric itself. Place and route tools for FPGA are completely different at the basic code level than those for ASICs.

There were some early attempts by EDA companies to make design software for FPGAs, but as the market came to be dominated by Xilinx and Altera, who provide their own tools, the economics of providing FPGA design tools became unattractive. In addition, the price point at which designers expected to purchase FPGA tools were an order of magnitude lower than for IC design tools. Consequently the EDA industry has largely ignored FPGAs.

A minor exception is seen in the existence of synthesis and floorplanning tools for high-end FPGAs, made by companies like Mentor and Synplicity (acquired by Synopsys in 2008). But the market has remained small because high-end becomes mainstream over time, and the free tools from the FPGA vendors become "good enough."

FPGAs now also include silicon IP, just as in the non-FPGA world. Increasingly, creating a large system-on-chip involves more assembly of pre-designed blocks of IP than creating original design. As systems rely more on software, FPGAs take on the role of a specialized computer consisting of a processor, peripherals, and perhaps some accelerators to act as the "secret sauce." The economics of providing IP for FPGAs has also not been attractive until recently, so most of the IP, apart from some processors, has been created by the FPGA vendors themselves.

In the early days of IP use in FPGAs, the IP blocks were relatively simple. At that time, the great majority of design software and IP for FPGAs were provided by the FPGA vendors directly. FPGA design tools are still largely provided by the FPGA vendors and there is little economic incentive for 3rd parties to provide these design tools.

However, for IP targeting FPGAs it has become another story entirely. As the size of FPGAs has grown, the complexity of the IP that can be used in them has also grown—from MSI devices to large subsystem blocks such as Ethernet MACs and controllers, PCIe interface blocks, SDRAM and NAND Flash memory controllers, video and network-packet processors, motor controllers, and even entire microprocessors.

The two major FPGA vendors directly supply microprocessor IP tailored specifically to their FPGA hardware architectures, but they increasingly rely on 3rd parties for specialized IP for applications such as networking, video and image processing, graphics, motor control, and other complex functions. This FPGA-specific 3rd-party IP runs through the FPGA design tools' design and logic-synthesis flow just as it does when designing chips using ASIC and SoC EDA tools. As a result, there's now a growing industry for 3rd-party companies developing FPGA-specific IP.

The Future of FPGA

Innovation continues in FPGAs. Starting in 2011, Xilinx started using through-silicon vias (TSVs) to build 3-dimensional chips. These are actually called 2.5D because multiple individual die are stacked on a silicon substrate called an interposer. True 3D has all the die stacked directly on top of each other in a single package. A TSV is just what it sounds like, a metal plug (usually copper) that runs from the top of the chip through the entire wafer to the bottom where it attaches to the interposer. Xilinx actually had the first 2.5D design in volume production.

In 2012, FPGA was about a $4.5 billion business. Xilinx has about 50% market share at $2.2 billion. Altera is not far behind at $1.8 billion. Actel, now part of Microsemi, and Lattice ($0.3 billion) are also significant suppliers.

Over the years, many startups have attempted to create FPGA companies to compete with Xilinx and Altera. One barrier to the entrepreneurs is that they run into patent infringement. Xilinx alone has about 2,500 patents on FPGAs and related topics. Another challenge is getting access to the leading-edge process technology that you need to be competitive. Xilinx and Altera, in particular, have deep relationships with the foundries and are likely to be ahead of any startup by a full process node.

However, two notable FPGA startups—Achronix and Tabula—have potentially solved the "access to leading process technology" problem by having Intel as an investor. Intel will manufacture their parts in 22 nm, marking them as two of the first customers of Intel's nascent foundry business.

In Their Own Words: Xilinx

> As the first company to design and sell FPGAs, and still the largest, Xilinx has a storied history of innovation and all the trial and error involved in creating a new product and a new market. In this chapter, Steve Leibson, strategic marketing director at Xilinx, describes the company's role in the development of the FPGA, both from a technology and a business perspective.

While working at microprocessor pioneer Zilog in the early 1980s, an engineer named Ross Freeman conceived of a new logic circuit that was reprogrammable: a single piece of silicon that could meet the needs of all of those ASIC customers. At that time, there were dozens, perhaps hundreds, of ASIC companies building custom silicon

Ross Freeman, inventor of the FPGA.

for thousands of customers. However, the design and fabrication of ASICs took months. Freeman's idea would permit the development and implementation of a custom IC in less than a day and it coincidentally hastened the birth of the fabless semiconductor industry.

Freeman earned a BS degree in physics from Michigan State University in 1969 and a master's from the University of Illinois in 1971. He worked in the Peace Corps and taught math in Ghana for two years. When he returned to the United States, Freeman joined Teletype Corporation and designed a PMOS (p-type metal-oxide semiconductor) chip. Back then, PMOS calculator chips were the profitable, high-volume LSI chips to make and PMOS was the process technology of choice because PMOS logic was the easiest type of MOS logic to manufacture. Therefore, it was also the cheapest type of logic to make. Freeman then became one of the first engineers to join a new microprocessor startup named Zilog where he designed the Zilog Z80-SIO (Serial I/O) peripheral chip.

By the time he reached his early 30s, Freeman was the Director of Engineering for Zilog's Components Division. He first got the idea for a new type of hardware-programmable device while working at Zilog and filed several patents. However, Zilog was not interested in pursuing the concept so Freeman left Zilog to further develop his idea. The result came to be known as the FPGA.

Although he had yet to develop a hardware design for this new device, the invention was impressive enough for Freeman to sway a Zilog coworker, Jim Barnett, to join him. The two of them then set out to recruit their former boss at Zilog, an experienced executive named Bernie Vonderschmitt, to become the CEO of the new FPGA start-up.

Semiconductor Business Lessons at RCA

Prior to joining Zilog, Vonderschmitt had worked for more than three decades at RCA Corp. He was hand-picked by RCA's legendary leader, David Sarnoff, to head the color television development in 1953. Even though the FCC had already approved it, Sarnoff was determined to obsolete the rival CBS color TV system that was based on a mechanical, rotating color wheel before it could spread commercially.

In an aggressive project over an 18-month period, Vonderschmitt's team at RCA developed the NTSC transmission standard. Unlike the CBS version, it was backward-compatible with the existing standard for black-and-white broadcast TV signals. Although often referred to by TV engineers as "Never The Same Color," the resulting NTSC broadcast standard remained in use for half a century until it was finally replaced in the United States by digital HDTV and the ATSC broadcast standard in 2009.

Partly because of his success in managing the NTSC color TV project, Vonderschmitt eventually became the Vice President and General Manager of RCA's Solid State Division. RCA had developed semiconductors for its own use in TVs, broadcast equipment and computers. The company waited until the late 1950s before becoming a merchant semiconductor vendor. Consequently, RCA missed the first IC manufacturing wave and did not become a major player in the early bipolar integrated circuit market like Fairchild. Instead, RCA focused on MOS integrated circuits.

In the early to middle 1960s, RCA's David Sarnoff Research Center developed a way to put both P- and N-channel transistors on one chip—demonstrating feasibility in 1963 and 1964 and then developing the COSMOS (RCA's trade name for "complementary symmetry metal oxide semiconductor," better known as CMOS) line of integrated circuits in the late 1960s. Vonderschmitt became head of RCA's Solid State Division a bit later, in 1972. During that period, Seiko came to RCA's Solid State Division seeking to license a low-power semiconductor process technology to help leapfrog its wristwatch business. That's when Vonderschmitt and Seiko first connected.

Seiko Epson Corporation had its origins in Suwa Seikosha, one of several manufacturing companies in The Seiko Group. The Seiko Group evolved from K. Hattori & Company, a trading company first established in 1881 that imported and exported clocks and watches. Suwa Seikosha was the manufacturing arm of the company. It made men's watches. Seiko had foreseen the changeover from mechanical watches to all-electronic wristwatches and it wished to be on the forefront of that change. Vonderschmitt licensed RCA's CMOS process to Seiko. By 1973, the company was producing digital LCD wristwatches based on Seiko's proprietary CMOS watch chips.

While serving as head of RCA's Solid State Division, Vonderschmitt had a clear view of semiconductor manufacturing's voracious appetite for capital; he managed the company's IC development and manufacturing businesses, overseeing three in-house semiconductor foundries. During his tenure as division head, Vonderschmitt often had trouble obtaining capital from the parent corporation to scale new IC process technologies from the research labs to production.

Things got more difficult as RCA moved further into its conglomerate phase. David Sarnoff's son Robert Sarnoff had taken over the reins of the company in 1970 and RCA announced its termination of the general-purpose computer systems division in 1971, marking the company's initial move away from technology and the start of its conglomerate phase. Making ICs was never RCA's mainstream business. Producing televisions, broadcast equipment, TV shows, and audio recordings on vinyl LPs were RCA's main businesses. During this period, RCA acquired Hertz (rental cars), Banquet (frozen foods), Coronet (carpeting), Random House (publishing) and Gibson (greeting cards). Consequently, those businesses and RCA's M&A activities garnered the lion's share of the company's development capital leaving little for IC manufacturing growth and development.

"If I Ever Start a Semiconductor Company, it Will Be Fabless"

By the end of his tenure at RCA, Vonderschmitt was convinced that captive semiconductor fabs were too expensive and too burdensome. "If I ever start a semiconductor company, it will be fabless," he vowed. "We'll find partners who can do our manufacturing for us." With these insights and his deep industry connections, Vonderschmitt had the vision and star power that Freeman and Barnett would need to secure investors for a fabless semiconductor company in 1984.

Bernie Vonderschmitt, originator of the fabless semiconductor business model.

Vonderschmitt left RCA in 1979 and decided to take time off from the industry. He used that time off to earn an MBA from Rider University. Then he joined microprocessor pioneer Zilog in Silicon

Valley. However, Vonderschmitt joined Zilog after its startup days, just after it was acquired by Exxon, and he soon saw that Zilog had the same problems with getting capital for its fab and for improving its semiconductor process technology from Exxon that RCA's Solid State Division had with its parent. It was "déjà vu all over again," to quote Yogi Berra, and Vonderschmitt was ready for a move when Freeman and Barnett approached him. Vonderschmitt, Freeman, and Barnett officially founded Xilinx in February 1984.

The Birth of Xilinx and the Fabless Movement

Although Freeman and Barnett convinced Vonderschmitt to found a new semiconductor company based on the FPGA's potential, he had no intention of owning fabs as did RCA or Zilog. Having twice experienced the stresses and risks of owning an IC fab, Vonderschmitt planned to focus Xilinx on what Xilinx did best—designing innovative programmable devices—and to partner with others to gain access to skills and assets not within Xilinx's area of expertise, especially capital-intensive chip manufacturing.

Bill Carter, designer of the first FPGA.

In pursuit of his vision for a fabless semiconductor company, Vonderschmitt now leveraged his decade-long friendship with a Seiko executive named Saburo Kusama to see if Seiko would be willing to manufacture FPGAs for Xilinx. Vonderschmitt had met Kusama-san while licensing RCA CMOS technology to Seiko for its watch business.

In pitching the Xilinx proposal, Vonderschmitt convincingly argued that such a partnership would let Seiko keep its fab running at capacity to further offset its capital equipment costs and possibly even make additional profit if the Xilinx FPGAs succeeded in the market. Vonderschmitt sweetened the deal by granting Seiko exclusive reselling rights to Xilinx FPGAs in Japan. The deal was consummated largely on the basis of the friendship between Vonderschmitt and Kusama-san. The initial paper agreement was only two pages long. Xilinx's fabless business was launched.

The task of actually designing the first functional FPGA fell to a young engineer named Bill Carter, whom Freeman and Barnett recruited from Zilog in March 1984. Freeman had originally hired Carter at Zilog to work on the Z8000 microprocessor. Carter now followed Freeman to Xilinx.

Prior to joining Xilinx, Carter had worked on NMOS microprocessors and peripherals at Zilog and also had previous bipolar design experience. But Seiko's process technology was CMOS, so the Xilinx FPGA would be Carter's first CMOS design. As an added challenge, this FPGA was going to be a very large chip. Thus Carter had to figure out a way to develop a never-designed-before IC, which would be as large as a complex microprocessor, on a very tight schedule. He also had to work with an IC fab on the opposite side of the Pacific that was not accustomed to working with outside customers all while overcoming barriers of a foreign language, foreign business culture, and a foreign engineering culture.

Nothing Too Clever or Exotic

Vonderschmitt regularly advised Carter to keep the design as simple as possible and not try anything "too clever or exotic." An overly complex design could make it harder to produce a functional device and deliver it to customers on schedule. Keeping risk to a minimum was very important to Vonderschmitt. He realized that a tiny start-up company offering a first-of-its-kind chip through a unique fabless business model could easily scare off customers.

In fact, to downplay the risk of doing business with the new FPGA company, Vonderschmitt told would-be customers that Xilinx planned to build a fab once it hit a US $50 million run rate and would also secure a second source, as was customary—actually almost mandatory—at the time. Monolithic Memories (MMI) later signed on as Xilinx's first second source. By coincidence, MMI was subsequently acquired by AMD, run by Jerry "real men have fabs" Sanders, and so AMD became a second source for Xilinx FPGAs.

Seiko's CMOS fab employed a 2.5 μm process—a relatively mature and low-risk silicon process well suited to digital watch circuits. Consequently, Seiko imposed very conservative design rules to make chip manufacturing easier and to boost yields. The Xilinx FPGA design would not be conservative with respect to design spacing. In fact, the XC2064 FPGA would require a whopping 85,000 transistors for its 64 configurable logic blocks and 58 input/output (I/O) blocks. That was more transistors than used in the design of the Motorola 68000 32-bit microprocessor. At roughly 300 millimeters (mm) per side, the die size for the first Xilinx FPGA would be larger than almost anything being manufactured at that time and it would be much bigger than anything Seiko's own designers had ever attempted.

To keep the FPGA within the 300-mm spec, Carter knew he would have to pack everything as close together as possible. He pushed Seiko to thoroughly characterize its CMOS process and to provide extremely accurate minimum feature sizes.

The chip's architecture was largely based on one modular CLB (configurable logic block) and one modular I/O block. The chip's design repeated these two blocks many, many times. (Some of the edge and corner CLBs required slight variations.) The repetitive use of identical modular blocks greatly simplified the FPGA's design—enough to permit manual design and verification. Xilinx used no computer-aided design (CAD). CAD systems were too expensive for a semiconductor start-up on a shoestring budget.

Carter's design team employed extensive design reuse so that it could concentrate the bulk of its time on circuit-level design and verification. Despite Vonderschmitt's urging to keep things simple, some of the chip-design techniques that Carter used were unconventional. For example, a typical CMOS design always pairs one p-channel transistor with an n-channel device. Carter's FPGA design drew on his NMOS design experience and employed fewer p-channel and more n-channel devices, which improved performance and saved space on the chip.

The design budget did allow Carter's design team to lease SPICE simulation time on a Control Data Corporation (CDC) mainframe, accessed via a dial-up connection in those pre-Internet days. Remote SPICE simulation was extremely slow. A simple syntactic error or typo could mean many lost hours and waiting in the timeshare queue for the SPICE job to run only to have that run fail because of a silly mistake was extremely frustrating, particularly with a looming delivery deadline.

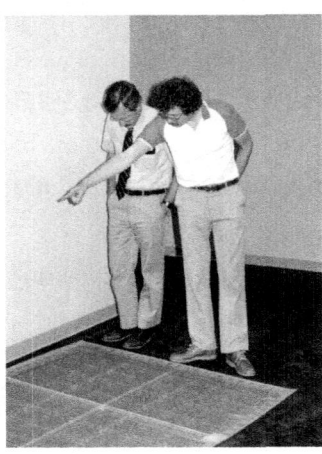

Ross Freeman checks a photoplot for the world's first FPGA, the XC2064.

Luckily, an inexpensive PC-based SPICE simulator became available at just the right time. Carter convinced Vonderschmitt to invest in a PC, which he used to verify that the SPICE deck syntax was correct before submitting the simulation to the CDC mainframe. Although the PC ran the SPICE simulation very slowly compared to the timeshare mainframe, the elapsed time was about the same due to the additional overhead from the upload speed over a modem and the waiting time in the timeshare queue. The CDC timeshare subscription quickly went away.

The team performed all design-rule checks, including electronic CAD (ECAD) electrical rule checks to find simple errors, at the end of the design process and then manually typed the final layout's cell coordinates into a Calma digitizer, which allowed the team to finally see the complete design layout for the first time. After further checks, the team sent the nine-layer design out to a pattern generator in preparation for mask production, which was done by Seiko. The chip taped out in late May 1985.

The First FPGA Wafers Were Mostly DOA

After delivering the design to Seiko, Carter's team had to wait two months until early July, 1985 to receive the first-run silicon: a box of 25 wafers. The team applied probes, a home-brew debugger, and a curve tracer to the wafers to see if any of the chips on the wafers would power up. The first ten wafers out of the box all exhibited dead shorts between power and ground. Not a good sign.

The 11th wafer showed some signs of life, and a very high current draw. The last 14 wafers also had dead shorts between power and ground. Carter's team discovered that insufficient metal etching was causing the shorts. There were aluminum whiskers causing shorts between the power and ground rails all over the first-run wafers. On the single partly dead wafer, the metal whiskers were thin enough to blow out like fuses. By applying enough supply current to a chip on the wafer, the test team managed to burn out the whiskers and open the shorts.

The now-operational chips on that one wafer were sufficiently robust to let Carter's team continue debugging the design of the FPGA. Carter was finally able to run a simple configuration bit stream into the device. After successfully programming an inverter into one of the CLBs, Carter called Freeman and Vonderschmitt while they were traveling in Japan and reported that that the "DONE line had gone high," and that Xilinx "had successfully created the world's most expensive inverter." After this initial success, the design team was able to program more and different logic circuits into the FPGA, eventually configuring all 64 CLBs on a chip.

Die Photo of the XC2064, the world's first FPGA.

The Birth of the FPGA

Seiko and Xilinx solved the aluminum-whisker issue that plagued the first-run batch of FPGA wafers and Xilinx received working devices in September, 1985. A press release announced the birth of the world's first FPGA—the XC2064 (called a "Logic Cell Array" in the press release)—on November 1, 1985.

In addition to providing an additional revenue stream to Seiko and helping to keep the fab full, running FPGAs with their repetitive structures through its fabs helped Seiko debug subsequent IC process technology generations, improving device yields and lowering unit costs for all of the chips Seiko manufactured. This use of FPGAs as a "process driver" helped to open the door for Xilinx, giving the company access to leading-edge IC manufacturing technology at Seiko and at other semiconductor foundries. Fab vendors now want to use FPGAs as process drivers because of their ability to diagnose process problems.

Conclusion

As the value proposition for the silicon foundry business became clear, other IDMs started filling their fabs and supplementing their revenue by manufacturing chips for third parties. In due course, an entirely new sub-industry emerged of dedicated, merchant semiconductor foundries that serviced fabless semiconductor vendors. This allowed even a small company of entrepreneurial designers to realize their innovations in silicon without the need to invest in a fab. The fabless revolution soon jumped into high gear and in 1994, ten years after Xilinx was founded, several fabless semiconductor companies including Xilinx formed the FSA (the Fabless Semiconductor Association, now called the GSA or Global Semiconductor Alliance) to provide a common voice to the electronics industry.

Bernie Vonderschmitt had foreseen all of these changes back in the 1970s while working for RCA. He knew from experience that fabs need more than one corporate customer to keep the fab lines filled and the lights on. He also understood that companies focused on IC design don't need to and often cannot afford to divert energy, resources, and attention to keeping their fab processes at the leading edge.

As we've ridden Moore's Law into nanometer territory through myriad tectonic IC process changes (copper interconnect, immersion lithography, high-K dielectric with metal gates, stress engineering, etc.), Vonderschmitt's fabless vision has become truer than ever. Over nearly thirty years, Xilinx has worked with more than 20 different semiconductor vendors and has partnered with ten vendors to supply its production FPGA silicon.

Vonderschmitt's vision has sustained Xilinx through three decades of FPGA leadership. The following words, written by Vonderschmitt two decades ago and published in the Xilinx Xcell newsletter, are still as accurate as the day they were written:

Making Fabless Strategies Work, by Bernie Vonderschmitt (1993)

> Xilinx is just one of the more than 100 semiconductor companies that do not own their own fabrication facility, and use independent silicon "foundries" for fabrication services. "Fabless" companies are not a fad, their streamlined structure fits today's tumultuous, fast-moving marketplace. Being fabless allows Xilinx to concentrate on what we do best—the design and marketing of programmable logic devices.
>
> Hewlett-Packard's announcement that it is quitting the foundry business and the recent troubles of a few fabless companies have led some industry pundits to once again question the viability of fabless semiconductor suppliers. (Hewlett-Packard is not one of Xilinx's wafer sources.) We strongly believe that the oracles predicting the demise of the fabless semiconductor are wrong. While ours is not the best business model for everyone, Xilinx and many other fabless companies will continue to succeed by establishing a win-win business relationship with our foundry partners.
>
> The first key to a successful fabless strategy is to employ standard fabrication processes that are compatible with a variety of foundries. Xilinx FPGAs and EPLDs are based on "plain vanilla" SRAM and EPROM technologies. This allows us to benefit automatically from the industry's latest process improvements and to establish multiple foundry sources for our products.
>
> Multiple foundry sources ensure adequate and continuous product availability in case of disasters. Competition between foundries, as well as ongoing process and product

improvements ensure that price projections are met. In contrast, fabless companies with specialized processes have fewer potential suppliers and less leverage in the "foundry market." If a foundry agrees to a specialized process, prices inevitably will be higher due to the special attention needed to get and keep that process under control.

The relationship between a fabless semiconductor company and its foundries must be long term, based on mutual trust, and of benefit to both parties. Xilinx benefits financially by gaining access to advanced fabrication processes without huge capital investments. We can focus on innovation, providing value through the development of better products.

Our foundry partners benefit from diversifying their manufacturing capacity over different equipment markets; through Xilinx, they have gained access to a significant new market segment without incurring the expense of product and market development. Foundries minimize demand volatility through this market diversification. Assuming a long-term relationship, the foundry can improve its competitiveness compared to other manufacturers.

Xilinx's foundries have gained a significant additional benefit—the ability to use FPGAs as process "drivers"—the technology used to drive and verify process advancements. The regularity and 100% testability of our FPGAs facilitates defect analysis and fault testing.

Our foundries have learned that by applying 10% to 20% of their capacity to FPGAs, they gain excellent rewards from process control diagnostics. The resulting improvements to their processes can be applied to their other CMOS product lines. (It should also be noted that Xilinx employs its own process experts, who work closely with our foundry partners in the development and implementation of process technology improvements.)

Thus, Xilinx can effectively drive process improvements through our working relationships with our foundry partners. But these relationships must be based on mutual benefits. In the future, as in the past, this will be a necessary ingredient for success.

Chapter 4: Moving to the Fabless Model

Before the mid-1980s, all semiconductor companies were what we now call Integrated Device Manufacturers or IDMs. They developed their own semiconductor process, purchased and ran their own fabs, and then sold the resulting product.

Fabs were never cheap, but back then a fab was affordable to even relatively small semiconductor companies. Aside from the capital costs of the manufacturing equipment, there are two other costs associated with having a fab: developing the manufacturing process and maintaining the volume required for cost-effective manufacturing. In the early 1980s, developing the semiconductor process was also affordable, and the volume required for cost-effective manufacturing was not too high, so "filling the fab" wasn't difficult.

Over the next couple of decades, all three expenses associated with running a fab changed. The cost to build a modern fab is now in the $10 BILLION dollar range. The process development cost for a modern semiconductor process is so high that only Intel and TSMC go it alone today. Everyone else is in a semiconductor company club of some sort where many of the costs of development are shared. For example, IBM, Samsung, and GLOBALFOUNDRIES together formed the Common Platform to collaboratively develop process technologies and populate the vast ecosystem of EDA, IP, libraries, packaging, and design services needed to bring a new design into existence.

Given the cost of building a modern fab and developing new process technology, the fab production line needs to be kept full. The volume required for cost-effective manufacturing runs into tens of thousands of wafer starts per week, a number so high that very few semiconductor companies can fill their own fab even if they could afford to build it. So today, IDMs are an endangered species and most semiconductor companies use foundries such as TSMC for all or most of their manufacturing. The fabless and fab-lite models are probably the single most important development since the invention of the integrated circuit in terms of its impact on the growth of the semiconductor industry.

Early Fabless Companies

This fabless trend started in 1984 with the first true fabless semiconductor companies—Chips and Technologies and Xilinx. Gordon Campbell, the co-founder and CEO of Chips and Technologies, is usually credited with realizing that a smaller semiconductor company could thrive without its own fab. Apparently, as Campbell and the other founders were raising funding for Chips and Technologies, the business plan included building a fab, although they never intended to do so. They believed no one would fund something as outrageous as a semiconductor company with no fab. They also suspected that the fabless idea was so appealing that once it was revealed, every venture capitalist they talked to would also start funding competitors because the investment required was so much lower than for an IDM.

Xilinx, founded in 1984, was a key early fabless semiconductor company. Xilinx was in a completely different business from Chips and Technologies, who built graphics chips for PCs, and they were not competitors. But their business model was basically the same; design a small number of products and pay another semiconductor company to manufacture the wafers. Without the large fixed investment of its own fab, the amount of investment required to get to market was much lower.

Another important early fabless semiconductor company was Qualcomm. Founded in 1985, it was involved in a number of different businesses, such as satellite location tracking used by long-haul truckers. In 1990, they developed the CDMA wireless standard, which was deployed in the US by Sprint and Verizon and grew their business very quickly.

When Chips and Technologies, Xilinx, and Qualcomm were founded, there were no pure-play foundries, companies that focus only on the manufacturing of other people's chips. However, semiconductor companies with their own fabs regularly manufactured wafers for other companies to even out an irregular production schedule. Any given semiconductor company might need more wafers than their fab could handle in one quarter, then have extra capacity the next quarter. Because so much of the cost of a fab is depreciation on the equipment, idle production lines cost almost as much as busy ones, in the same way that an empty airline seat costs almost as much as filling the seat.

A big advantage of the fabless model was shedding the high fixed cost of running a fab. In effect, the fixed costs of a fab were changed to variable costs. Fabless companies paid a little more per wafer by using someone else's fab, but this was offset by the money saved by not having to build, run, and fill their own fab.

It was into this burgeoning new fabless semiconductor industry that the first pure-play foundry, TSMC, was launched in 1987. It was pure-play in the sense that it only manufactured and sold wafers for other companies. It did not design any of its own products. Across the road from TSMC was United Microelectronics Corporation, or UMC, which was founded as a traditional semiconductor company (and also provided foundry services) in 1980. UMC became a pure-play foundry in 1995, and has ranked second to TSMC until recently. The next chapter focuses on the foundry business in more depth.

In the early days, TSMC and UMC mostly manufactured wafers for semiconductor companies who had fabs but not enough capacity. There was very little business from fabless companies. In fact, all the designs from fabless semiconductor companies together could not have filled a single fab in 1995. The balance would gradually change in favor of fabless semiconductor companies and was driven by two trends. First, the creation of more and more fabless semiconductor companies, which all required manufacturing capacity; second, the gradual decline in the economic attractiveness to semiconductor companies to own their own fabs.

When then CEO of AMD, Jerry Sanders, said "Real men have fabs," he was suggesting that chip design and process technology had to be tightly coupled. One could argue what impact this has on a business. AMD abandoned its fabs in 2009, and their main competitor, Intel, did not. For all the missteps AMD has made in the market, it's not clear that any of them are due to the decoupling of design and manufacturing that resulted when they went fabless. The world semiconductor rankings from iSuppli show the shift towards the fabless model. As recently as 2007, there were no fabless semiconductor companies ranked in the top ten. In 2012, the latest year for which figures are available, Qualcomm is number three and Broadcom, also fabless, is number nine, and at number 12, is the now-fabless AMD.

Today there is actually very little difference between the traditional IDMs and the fabless companies. Most semiconductor design houses outsource at least some of their manufacturing to foundries such as TSMC or GLOBALFOUNDRIES. State-of-the-art fabs are so expensive that even semiconductor companies that keep some of their manufacturing in-house are forced to use foundries for their most advanced manufacturing at 28 nm, 20 nm, and below. Only a few IDMs—such as Intel, IBM, and Samsung—have state-of-the-art fabs, although they also outsource some manufacturing to the foundries.

The history of pure-play foundries and their role in shaping today's fabless semiconductor industry is an interesting story in itself, and in fact, after the following history of fabless company Chips and Technologies, is the topic of the next chapter.

In Their Own Words: Chips and Technologies

> *Chips and Technologies (C&T) is known as the first fabless semiconductor company. One of the founders, Dado Banatao, spoke to author Paul McLellan about starting the company and launching the business model that would eventually be dominant in the industry.*

In 1985, Dado Banatao and Gordon Campbell started a company called Chips and Technologies. Campbell, along with Bernie Vonderschmitt of Xilinx, is credited with pioneering the new, fabless, business model. Disruptive new concepts usually meet with resistance by the establishment, and in fact Banatao, who today is managing partner at Tallwood Venture Capital, said that they had a hard time raising money because VCs couldn't comprehend a fabless semiconductor company. Even his friends told him it "wasn't a real semiconductor company." In fact the first $1 million was raised from a real-estate investor! Only after they were further along were they able to raise another $3 million from various Japanese investors, including the large Japanese conglomerate, Mitsui Group.

They made a technical decision to use gate-array, rather than standard-cell design, because the booming PC business created a sense of urgency in getting their chipsets to market. They chose the industry-leading gate-array technology from Toshiba, but then found that their design was too large for even the biggest gate-array. The solution was to partition the design into two chips: one logic CMOS gate-array from Toshiba and a separate bipolar chip with all the IO drivers that Hitachi fabricated. Hitachi had a lot of available fab capacity due to the semiconductor downturn at the time and was somewhat desperate for business. C&T filled up those fab lines completely, and got unbelievably low prices.

C&T's business took off fast. C&T sold chipsets for IBM's PC-XT and PC-AT, and made $12 million profit in the first four months after their chipset was introduced. By the time they had their IPO in 1987, just 22 months after opening, they still had $1M of their original $4 million investment in the bank. The fact that Mitsui was an investor turned out to be fortuitous because Toshiba and Hitachi were part of the Mitsui Group. This let them order parts from Toshiba and Hitachi without having to pay up-front with working capital that they didn't have. Mitsui financed $50 million in inventory.

C&T products dominated for about two years before their competitor, VLSI Technologies, started selling their first chipset. C&T's bipolar chip turned out to have the advantage, since at that time electrostatic discharge (ESD) protection on the CMOS technology favored by VLSI was in its infancy and was still a potential problem. This let C&T make the claim of better reliability compared to VLSI's all-CMOS solution. Three years later C&T's chipset was all CMOS too, but by then ESD protection was up to 20 KV and those issues had gone away.

Banatao left C&T in 1989 to launch another start-up, S3 Graphics, which focused on graphics processors. S3 Graphics also used the gate-array technology to get to market fast. They looked around for the biggest arrays and found what they needed at Seiko-Epson. Banatao's key invention at S3 was a new interconnect, a local bus, to move between chips faster. They called it Advanced Chip Interconnect, which later became Intel's PCI and PCIe.

C&T also sold chipsets to Dell very early on and they rode the PC rocket together. While Compaq was the top PC maker at the time, they didn't yet believe in using chipsets. Before long, Taiwan, Korea, and Japan were all making PCs. Compaq couldn't compete, so they switched from single chips to chipsets too. Interestingly, at that point in the industry history, C&T was making more on each PC than Intel was.

Gordon Campbell, who had come from Intel and had a previous startup called SEEQ Technology, left the C&T in 1993 to found 3Dfx Interactive and has been involved with many other successful companies since then. By the time he left C&T, the company had been suffering from lower sales and big losses. The new president and chief executive officer, Jim Stafford, managed to get them back to profitability by restructuring the management and some product lines.

C&T went on to become one of the largest suppliers of graphics processors for notebook PCs. They began to work with Intel and Lockheed Martin on a new graphics chip for desktop PCs and workstations, the Intel740, which was unveiled in early 1997. This collaboration proved to be more than was evident at the time. By July 1997, Intel's all cash offer of about $400 million to acquire C&T was a done deal. This was Intel's largest acquisition ever.

The impact of the new fabless business model started by C&T can't be understated. This is by far the dominant model for the semiconductor industry today.

Chapter 5: The Rise of the Foundry

The fundamental economics of the semiconductor industry are summed up in the phrase "fill the fab." Building a fab is a major investment. With a lifetime of just a few years, the costs of owning a fab are dominated by depreciation of the fixed capital assets (the building, the air and water purification equipment, the manufacturing equipment, etc.). This puts a big premium on filling the fab and running it as close to capacity as possible. If a fab is not full, then the fixed costs will overwhelm the profit on the capacity that is used and the fab will lose money. Of course, if demand is high there is a corresponding problem because a fab that is already full cannot, by definition, manufacture any more.

The capacity of a fab usually meets most of the overall needs of the company that built it, but there can be a mismatch between the capacity needed, in terms of wafer-starts, and what is available. Sometimes, the fab is full, but the semiconductor company could sell more product if only they could have manufactured it. Other times, the fab has surplus capacity, and the semiconductor company doesn't have enough product to keep the fab full. This need to balance fab capacity and demand led to the original foundry businesses, in which semiconductor companies, even competitors, bought raw manufactured wafers from each other.

The first fabless semiconductor companies, such as Chips and Technologies and Xilinx, extended this model a little bit. By definition, they didn't have their own fabs, but they would form strategic relationships with semiconductor companies that had excess capacity. The relationships had to be strategic: you couldn't just walk into a semiconductor company and ask for a price for a few thousand wafers, any more than today you can walk into, say, Ford and ask how much to have a few thousand cars manufactured. It is not how they are set up to do business.

Along Comes the Pure-Play Foundries

In 1987, a major change in the semiconductor industry took place with the creation of TSMC. It was an outgrowth of Taiwan's Industrial Technology Research Institute, ITRI. Because very few fabless semiconductor companies existed back then (Chips and Technologies was founded only in 1985 for instance), the TSMC business model was to provide manufacturing services to semiconductor companies who were short of capacity in their own fabs. One of the original investors in TSMC was Philips Electronics (since spun-out from Philips as NXP), who was also one of the first customers buying wafers from them.

UMC, was an earlier spinoff from ITRI, created in 1980 as Taiwan's first semiconductor company. Across the road from TSMC in Hsinchu, its focus also gradually shifted to foundry manufacturing, especially once the fabless ecosystem created both a lot of demand and a wish to have a competitor to TSMC to ensure that pricing remained competitive.

The third of the big three pure-play foundries back in that era was Chartered Semiconductor. Chartered was based in Singapore and backed by a consortium that included the Singapore government, who saw semiconductor manufacturing as a strategic move up the electronic value chain from contract manufacturing.

With the creation of TSMC, it became possible for semiconductor companies to have wafers manufactured without requiring a deep strategic relationship. Pricing wasn't so transparent that you could just look at the price-list on the web (plus, in 1987 there wasn't a web) but a salesman would quote you for whatever you needed. It was very similar to a metal foundry, where the name had come from; if you wanted some metal parts forged, the foundry would give you a quote and build them for you. In the same way, if you needed some wafers manufactured you could simply go and get a price.

This might not seem like that significant a change, but it meant that forming a fabless semiconductor company no longer depended on the founders of the company having a huge amount of capital to build a fab or some sort of inside track with a semiconductor company with a fab. They could focus on doing their design safe in the knowledge that when they reached the manufacturing stage, they could simply buy wafers from TSMC, UMC, or other companies that had entered the foundry business.

Companies such as TSMC and UMC were known as pure-play foundries because they didn't have any other significant lines of business. Semiconductor companies with surplus capacity would still sell wafers and run their own foundry businesses, but they were always regarded as a little bit unreliable. Everyone suspected that if the semiconductor company's business expanded, then their fab would close to outsiders, forcing the companies using that fab to find a new supplier. Gradually, over time, the semiconductor companies whose primary business was making their own chips became known as IDMs. This is in contrast to the fabless ecosystem, in which the companies that create and sell the designs (the fabless semiconductor companies), are separate from the companies that manufacture them (the foundries).

Foundries Drive the Transformation from IDM to Fabless

The line between fabless semiconductor companies and IDMs has blurred over the last decade. Back in the 1990s, most IDMs manufactured most of their own product, perhaps using a foundry for a small percentage of additional capacity when required. But their own manufacturing was competitive, both in terms of the capacity of fab they could afford to build, and in terms of process technology.

Gradually, both of these things changed. The size of fab required to remain cost-competitive continued to increase to the point that most semiconductor companies could not fill a fab that large. In 2002, the CEO of Intel, Paul Otellini, estimated the cost of building a new fab at

$2 billion. Samsung spent $4 billion to build a fab in 2006. Estimates in 2013 for new fabs in the planning stages for GLOBALFOUNDRIES and Intel hover around $10 billion. The semiconductor processes also got significantly more complex and costly, so that the cost of staying on the leading edge became prohibitive for all except the largest IDMs, most notably Intel, IBM, and Samsung.

The first key change away from the dominance of IDMs was the formation of several process clubs, cooperative deals in which much of the cost of semiconductor process technology development is shared between a number of semiconductor companies. One early processor club was the 1992 alliance between IBM, Toshiba, and Siemens to develop memory chips. A small semiconductor company couldn't hope to develop a state-of-the-art process on its own.

It quickly became clear that only the largest semiconductor companies could afford to build a cost-competitive fab. It wasn't just a matter of the investment required, but also that there would be more capacity than they would be able to use. Back when a fab cost $3 billion to build, a company would face a depreciation cost of roughly $1 billion per year, meaning that they need to have a running semiconductor business of perhaps $5 billion, around the size of AMD, who was the only competitor to Intel in the x86 microprocessor business.

In fact, in March 2009, AMD, who's CEO was famous for the "Real men have fabs" comment, went completely fabless. They divested their manufacturing to the Advanced Technology Investment Company, primarily owned by the Emirate of Abu Dhabi. The manufacturing part became the pure-play foundry GLOBALFOUNDRIES, although AMD maintained a stake in the new foundry and was its largest customer. Subsequently, GLOBALFOUNDRIES acquired Chartered Semiconductor and today it is the second largest foundry behind TSMC.

Many other semiconductor companies also went fabless, such as Freescale, Infineon, and Sony. Other semiconductor companies didn't go quite so far. They kept their existing fabs, many of which were fully depreciated and running non-leading-edge processes. But for the most advanced processes, they used the pure-play foundries because they couldn't afford either the investment or the cost of technology development to keep up.

In the meantime, some IDMs also entered the foundry business. They are not pure-play foundries since they also have a merchant semiconductor business as well, but they are leveraging their leading edge process capability by selling wafers too. The three most significant of these IDMs with a foundry business are Samsung, IBM, and Intel.

Samsung is Apple's largest semiconductor supplier despite also being their leading competitor in the smartphone market. It is also true the other way around; Apple is Samsung's largest foundry customer. In 2012, Samsung overtook UMC to become the 3rd largest foundry.

IBM consumes most of their product internally to build computer systems but also does a certain amount of foundry business, such as building the logic chip for Micron's hybrid memory cube.

Intel has also entered the foundry business. Initially, most of its customers have been in the FPGA business, most notably Altera (the second biggest FPGA company), which is using Intel's 14 nm process. Intel has also invested in, and is doing the manufacturing for, a few FPGA startups such as Achronix and Tabula. However, Microsemi (which purchased Actel, another FPGA company) also selected Intel as a foundry and is using their 22 nm process. There have also been reports that Cisco is using Intel as a foundry.

Today, the foundry model has split into two separate businesses. At the leading edge, currently 22/20 nm and 14/16 nm, there are just a handful of foundries with the financial and commercial muscle to build a state-of-the-art fab. For pure-play foundries, there is TSMC and GLOBALFOUNDRIES, and UMC with more limited capacity. TSMC has fabs in Taiwan for these nodes, and GLOBALFOUNDRIES is running 28 nm in Dresden and building a brand new fab in Malta, New York. These are the foundries that can build the latest smartphone SoCs such as those by Apple, nVidia, and Qualcomm.

The other leading edge foundries are the IDMs with a foundry business: Intel, with fabs in Oregon, Arizona, and Ireland (and many more fabs in other places), and Samsung, with fabs in Korea and Texas. IBM has its leading edge fab in East Fishkill, New York.

The transition from IDMs to the foundry model has been quite dramatic. At 130 nm there were 22 IDMs with their own fabs. By 45 nm this was down to nine IDMs and five foundries. At 22 nm there is only Intel, Samsung, and IBM as IDMs along with TSMC, GLOBALFOUNDRIES, UMC, perhaps SMIC, and Samsung as foundries. At 14/16 nm the list looks like it may shrink again. These companies are the only ones that have announced fabs manufacturing at process nodes below 20 nm. Every other semiconductor company is either fabless or fab-lite for these leading-edge process nodes. The adoption of these advanced nodes is driven by microprocessors (for Intel) and smartphones (for everyone else), very high volume businesses that need the highest possible performance and the lowest possible power.

The other part of the foundry business is not focused on the leading edge but on designs for analog, power management, micromechanical, and so on. For these designs, the state-of-the-art process today is 130 nm and so does not require the most leading-edge fab. The Chartered fabs that GLOBALFOUNDRIES acquired run this type of business, as do some other specialized fabs such as Tower/Jazz and Vanguard, both of which are pure-play foundries. There are also some IDMs running this sort of business, such as PowerChip and MagnaChip.

Foundries Grow, Fabless Grows

Data from the GSA show that in 2013, the top foundry is TSMC by a long way, with 2013 revenues of $20.1 billion. GLOBALFOUNDRIES is number two with revenues of $5.1 billion. The third largest company in terms of foundry business is Samsung (which is an IDM) at $4.6 billion, followed by UMC with $3.965 billion. Fifth on the list is China-based SMIC, the only other company with foundry revenues over $1 billion, at $1.970 billion. The size of business drops off rapidly although there is a long tail of foundries. For example, in twelfth place is Korea-based MagnaChip with revenues of $440 million, less than one fortieth the size of TSMC.

A forecast from IC Insights published in the GSA 2014 *IC Foundry Almanac* shows foundry revenues (both pure-play and IDMs) are expected to increase 15% in 2014 to a record-high $51.1 billion. This follows 14% growth in 2013 and 21% growth in 2012. Foundries are now responsible for more than one third of IC sales worldwide. By 2017, foundry manufactured ICs are expected to represent 45% of the industry's total integrated circuit revenues.

More and more of the top ten non-memory semiconductor manufacturers are fabless and fab-lite companies. In iSuppli's 2013 preliminary rankings Intel (IDM), Samsung (IDM and foundry), and Qualcomm (fabless) take up the top three places. They are followed by Micron (memory), SK Hynix (memory), Toshiba (fab-lite), Texas Instruments (fab-lite), Broadcom (fabless), ST Microelectronics (fab-lite) and Renasas (fab-lite). Basically, the rest of the top 20 are all either completely fabless, take the fab-lite approach of using foundries for leading edge process while internally manufacturing anything that doesn't require a leading-edge fab, or they are specialized analog suppliers who don't use leading edge processes at all. IBM (IDM and foundry) slots in somewhere, but they consume so much of their silicon internally that it is not clear where. Their merchant business is not very large.

In the future, it is unclear whether all the IDMs and foundries that have made the move to 22 nm will be able to afford to make the transition to 14/16 nm, and subsequently to 10 nm and 7 nm. There are big technical challenges as well as economic issues as to what wafers will cost. As a result, just how much of the existing product lines will make the transition to new process nodes, as opposed to remaining on cheaper, less advanced, process nodes is unclear.

An additional wrinkle is that it is not clear whether the cost per million transistors—a key measure of the value of migrating to smaller process nodes—will continue to decline as it has for the past 40 years. One view of the world is that 28 nm might be the cheapest process by this metric. Of course, at 20/22 nm and 14/16 nm there are gains in performance and reductions in power, but the cost reduction that has always accompanied a process transition may not materialize, or at the very least will be significantly reduced. The old rule of thumb for a process transition was that you get twice as many transistors at a 15% increase in wafer cost, meaning a cost reduction of 35%. It remains to be seen how costs will change going forward.

One technology that the industry has hoped would allow Moore's Law to continue is EUV lithography. It has been delayed for years, and while the industry has invested heavily in the technology, it still cannot achieve high enough wafer throughput to be used for high-volume manufacturing, and it remains unclear if and when it will be. However, it also offers the opportunity for a significant decrease in the time required to manufacture wafers and a corresponding reduction in costs.

In the next two chapters, TSMC and GLOBALFOUNDRIES describe their histories and roles in development of the foundry business model.

In Their Own Words: TSMC and Open Innovation Platform

TSMC, the largest and most influential pure-play foundry, has many fascinating stories to tell. In this section, TSMC covers some of their basic history, and explains how creating an ecosystem of partners has been key to their success, and to the growth of the semiconductor industry.

The history of TSMC and its Open Innovation Platform (OIP)® is, like almost everything in semiconductors, driven by the economics of semiconductor manufacturing. Of course ICs started 50 years ago at Fairchild (very close to where Google is headquartered today, these things go in circles). The planarization approach, whereby a wafer (just 1" originally) went through each process step as a whole, led to mass production. Other companies such as Intel, National, Texas Instruments and AMD soon followed and started the era of the Integrated Device Manufacturer (although we didn't call them that back then, we just called them semiconductor companies).

The next step was the invention of ASICs with LSI Logic and VLSI Technology as the pioneers. This was the first step of separating design from manufacturing. Although the physical design was still done by the semiconductor company, the concept was executed by the system company. Perhaps the most important aspect of this change was not that part of the design was done at the system company, but rather the idea for the design and the responsibility for using it to build a successful business rested with the system company, whereas IDMs still had the "if we build it they will come" approach, with a catalog of standard parts.

In 1987, TSMC was founded and the separation between manufacture and design was complete. One missing piece of the puzzle was good physical design tools and Cadence was created in 1988 from the merger of SDA and ECAD (and soon after, Tangent). Cadence was the only supplier of design tools for physical place and route at the time. It was now possible for a system company to buy design tools, design their own chip and have TSMC manufacture it. The system company was completely responsible for the concept, the design, and selling the end-product (either the chip itself or a system containing it). TSMC was completely responsible for the manufacturing (usually including test, packaging and logistics too).

At the time, the interface between the foundry and the design group was fairly simple. The foundry would produce design rules and SPICE parameters for the designers; the design would be given back to the foundry as a GDSII file and a test program. Basic standard cells were required, and these were available on the open market from companies like Artisan, or some groups would design their own. Eventually TSMC would supply standard cells, either designed in-house or from Artisan or other library vendors (bearing an underlining royalty model transparent to end users). However, as manufacturing complexity grew, the gap between manufacturing and design grew too. This caused a big problem for TSMC: there was a lag from when TSMC wanted to get designs into high volume manufacturing and when the design groups were ready to tape out. Since a huge part of the cost of a fab is depreciation on the building and the equipment, which is largely fixed, this was a problem that needed to be addressed.

At 65 nm TSMC started the Open Innovation Platform (OIP) program. It began at a relatively small scale but from 65 nm to 40 nm to 28 nm the amount of manpower involved went up by a factor of 7. By 16 nm FinFET, half of the design effort is IP qualification and physical design because IP is used so extensively in modern SoCs, OIP actively collaborated with EDA and IP vendors early in the life-cycle of each process to ensure that design flows and critical IP were ready early. In this way, designs would tape-out just in time as the fab was starting to ramp, so that the demand for wafers was well-matched with the supply.

In some ways the industry has gone a full circle, with the foundry and the design ecosystem together operating as a virtual IDM. The existence of TSMC's OIP program further sped up disaggregation of the semiconductor supply chain. Partly, this was enabled by the existence of a healthy EDA industry and an increasingly healthy IP industry. As chip designs had grown more complex and entered the SoC era, the amount of IP on each chip was beyond the capability or the desire of each design group to create. But, especially in a new process, EDA and IP qualification was a problem.

On the EDA side, each new process came with some new discontinuous requirements that required more than just expanding the capacity and speed of the tools to keep up with increasing design size. Strained silicon, high-K metal gate, double patterning and FinFETs each require new support in the tools and designs to drive the development and test of the innovative technology.

On the IP side, design groups increasingly wanted to focus all their efforts on parts of their chip that differentiated them from their competition, and not on re-designing standard interfaces. But that meant that IP companies needed to create the standard interfaces and have them validated in silicon much earlier than before.

The result of OIP has been to create an ecosystem of EDA and IP companies, along with TSMC's manufacturing, to speed up innovation everywhere. Because EDA and IP groups need to start work before everything about the process is ready and stable, the OIP ecosystem requires a high level of cooperation and trust.

When TSMC was founded in 1987, it really created two industries. The first, obviously, is the foundry industry that TSMC pioneered before others entered. The second was the fabless semiconductor companies that do not need to invest in fabs. This has been so successful that two of the top 10 semiconductor companies, Qualcomm and Broadcom, are fabless and all the top FPGA companies are fabless.

The foundry/fabless model largely replaced IDMs and ASIC. An ecosystem of co-operating specialist companies innovates fast. The old model of having process, design tools and IP all integrated under one roof has largely disappeared, along with the "not invented here" syndrome that slowed progress since ideas from outside the IDMs had a tough time penetrating. Even some of the earliest IDMs from the "Real men have fabs" era have gone "fab lite" and use foundries for some of their capacity, typically at the most advanced nodes.

Legendary TSMC Chairman Morris Chang's "Grand Alliance" is a business model innovation of which OIP is an important part, gathering all the significant players together to support customers—not just EDA and IP, but also equipment and materials suppliers, especially for high-end lithography.

Digging down another level into OIP, there are several important components that allow TSMC to coordinate the design ecosystem for their customers.

- EDA: the commercial design tool business flourished when designs got too large for hand-crafted approaches and most semiconductor companies realized they did not have the expertise or resources in-house to develop all their own tools. This was driven more strongly in the front-end with the invention of ASIC, especially gate-arrays; and then in the back end with the invention of foundries.

- IP: this used to be a niche business with a mixed reputation, but now is very important with companies like ARM, Imagination, CEVA, Cadence, and Synopsys, all carrying portfolios of important IP such as microprocessors, DDRx, Ethernet, flash memory and so on. In fact, large SoCs now contain over 50% and sometimes as much as 80% IP.

TSMC has well over 5,500 qualified IP blocks for customers.

- Services: design services and other value-chain services calibrated with TSMC process technology helps customers maximize efficiency and profit, getting designs into high volume production rapidly
- Investment: TSMC and its customers invest over $12 billion a year. TSMC and its OIP partners alone invest over $1.5 billion. On advanced lithography, TSMC has further invested $1.3 billion in ASML.

Processes are continuing to get more advanced and complex, and the size of a fab that is economical also continues to increase. This means that collaboration needs to increase as the only way to both keep costs in check and ensure that all the pieces required for a successful design are ready just when they are needed.

TSMC has been building an increasingly rich ecosystem for over 25 years and feedback from partners is that they see benefits sooner and more consistently than when dealing with other foundries. Success comes from integrating usage, business models, technology and the OIP ecosystem so that everyone succeeds. There are a lot of moving parts that all have to be ready. It is not possible to design a modern SoC without design tools, more and more SoCs involve more and more 3rd party IP, and, at the heart of it all, the process and the manufacturing ramp with its associated yield learning all needs to be in place at TSMC.

The proof is in the numbers. Fabless growth in 2013 is forecasted to be 9%, over twice the increase for the overall industry at 4%. Fabless has doubled in size as a percentage of the semiconductor market from 8% to 16% during a period when the growth in the overall semiconductor market has been unimpressive. TSMC's contribution to semiconductor revenue grew from 10% to 17% over the same period.

The OIP ecosystem has been a key pillar in enabling this sea change in the semiconductor industry.

Global Unichip Corporation

Another facet of TSMC is GUC, Global Unichip Corporation. It is a partially owned subsidiary and also an important partner, providing design services and allowing TSMC themselves to continue to be a pure-play foundry. GUC was founded in 1998 with 10 employees as what has come to be known as a "Design Service" company. It ramped fast and by 2000 it employed over 100 people.

The years between 2003 and 2010 were milestone years for GUC, representing a period of unprecedented growth. The era was marked by a strengthening of both business and technology relationships with the largest semiconductor foundry in the world, TSMC. That relationship set GUC on firm growth, bringing over the core of today's management and the business strategy that guides the company today.

In 2003, TSMC assumed an ownership stake in GUC. But the foundry leader's investment went far beyond financial investment. Part of its strategy to enhance the return on its investment was to move GUC to a global business strategy and put it on the road to being an advanced technology leader.

The technology model, and the business model that accompanied it, soon began to gain traction. Prior to 2003, much of GUC's business came from the consumer electronics companies who tended to utilize more mature technologies and were primarily located in Taiwan. With the installation of new management, and new business and technology models, business emphasis began to migrate to the more technically sophisticated networking and communications sectors that required more advanced technologies. In 2004, 100% of GUC's revenue was at the 0.13 μm technology node; by 2005, 5% of revenue came from the new 90 nm node and a year later, an additional 3% of revenue came from the emerging 65 nm node.

The impact of this trend was soon seen in the company finances. Revenue jumped from $20 million in 2002 to $27 million in 2003, $32 million in 2004, and a whopping $48 million in 2005—more than doubling the revenue over a four year period.

The year 2006 marked another major milestone. In the third quarter of that year, GUC became a publicly traded company when it offered its shares on the Taiwan Stock Exchange.

The operations side instituted a major focus on advanced technology. In 2007, the company developed an advanced technology digital design flow, followed shortly thereafter by a low power design flow. As a result, the company saw a large increase in the size of their designs, many with gate counts jumping exponentially. In the face of an industry-wide recession in 2009, GUC showed its confidence in the future by investing heavily in internal IP development, in particular, IP targeting the networking market segment.

This era of prosperity was reflected by a broad set of indices. Annual revenue in 2006 more than doubled that of 2005 ($48 million) at $103 million, then more than doubled again in 2007 to $216 million. In 2008, revenue jumped to $295 million before falling during the recession of 2009 to $252 million. In 2008, GUC saw a jump in revenue from advanced technology to 21 % and to 34 % in 2009, with 1 % of that coming from leading edge 40 nm products. Like all technology companies, GUC experienced a financial decline in 2009, with revenues dropping to $252 million.

But the company would rebound quickly in 2010, posting revenues of $327 million with advanced technologies accounting for 42% of that total. The year also proved auspicious. Driven by the recession to examine its business model, GUC would begin making a series of strategic decisions that would allow it to capitalize on a new era of semiconductor device design.

The company's growth as an innovative force in the semiconductor industry is also reflected in the number of new employees required to implement increasingly complex technologies. At the end of 2003, GUC employed 132 people, most of them in Taiwan. Three years later, that number had more than doubled to 287 and by the start of 2010, the company counted 484 employees, a number that has held relatively steady through 2013. Employee growth was fueled by expanded geographic growth. GUC opened its first international office when it established a subsidiary in North America (GUC N.A.) in February of 2004 and then opened its Japan office in June of 2005. Nearly three years later, in May of 2008, the company opened its third international office, GUC Europe, in Amsterdam, The Netherlands and one month later opened an office in Korea. GUC entered the fast-growing China market when it opened an office in one of that country's technology hubs near Shanghai in 2009.

Success in the semiconductor industry going forward is going to be heavily weighted by the ability to leverage industry's third party infrastructure that has now matured. Foundries are at the leading edge of the infrastructure, providing the most advanced process technologies, as well as specialized technologies to all comers. IP and chip design implementation are also being outsourced, cost effectively utilizing technology and financial resources.

It is in this new and exciting environment that GUC evolved the Flexible ASIC Model, which is designed to provide the most effective, efficient and flexible path to semiconductor innovation.

The Flexible ASIC Model is a response to both the business and technical challenges facing today's semiconductor companies. This model allows companies to allocate their resources more efficiently. It brings together design expertise, systems knowledge and manufacturing resources to efficiently drive delivery of the final packaged IC. The model's basic strategy is to spread design risks and to minimize IDM, fabless and OEM (Original Equipment Manufacturers) upfront semiconductor-related fiscal and human capital investments. The goal is to increase the efficiency of the entire value chain, from concept to delivery; to shorten all phases of the development cycle; and to ultimately increase device yield quality and reliability.

At the heart of the Flexible ASIC Model is integrated manufacturing. GUC has made a strategic choice to work exclusively with TSMC, the semiconductor industry's leading foundry service company. It is this relationship that plays an integral role in the company's ability to achieve early advanced technology access and match designs to manufacturing resources.

Spotlight: Dr. Morris Chang

Dell changed the way personal computers are manufactured and sold. Starbucks changed the amount we would pay for a cup of coffee. Ebay took the yard sale out of our yards. TSMC took the semiconductor manufacturing costs off our balance sheets and out of our capital investments.

It's hard to overstate the impact that Dr. Morris Chang, Founder, Chairman, and until-recently CEO of TSMC, has had on the industry. He has been influential as a leader in business model innovation, and has earned his company roughly 50% of the foundry market share.

Chang left his native China in 1949, moving to the US to attend Harvard University. He soon transferred to MIT as he followed his interest in technology. After earning his MS in 1953 from MIT's mechanical engineering graduate school, Morris went directly into the semiconductor industry at the process level with Sylvania Semiconductor and was quickly moved to management.

Chang moved to Texas Instruments in 1958, where he stayed for 25 years, rising to VP of the worldwide semiconductor business (and also earned a PhD in electrical engineering from Stanford in 1964). At TI, he worked on a four transistor project in which the manufacturing was done by IBM, thus engaging in one of the early semiconductor-foundry relationships. Also at TI, Chang developed a new model of semiconductor pricing that sacrificed early profits to gain market share and to achieve manufacturing yields that would lead to higher long-term profits.

Chang left TI in 1983 and did a short stint at General Instrument Corporation. He then moved to Taiwan to head the Industrial Technology Research Institute (ITRI), which led to the founding of TSMC.

Chang noticed in the early 1980s, while at TI and GI, that top engineers were leaving and forming their own semiconductor companies.

Unfortunately, the heavy capital requirement of semiconductor manufacturing was a gating factor. The cost back then was $5-10 million to start a semiconductor company without manufacturing and $50-100 million to start a semiconductor company with manufacturing. Some of these start-ups used excess capacity from IDMs but were subjected to uncertainties in foundry capacity and sometimes had to buy wafers from a competitor. Around this time, 1985, the first truly fabless startups, like Xilinx and Chips and Technologies, were launched and doing well.

It was in 1987, within this nascent fabless environment, that Chang started TSMC. Although TSMC started two process nodes behind where semiconductor manufacturers (IDMs) were at the time, they had the advantage of being a pure-play foundry, not a competitor. Their focus was on their customers.

Morris Chang made the first TSMC sales calls with a single brochure: TSMC Core Values: Integrity, Commitment, Innovation, Partnership. Four or five years later, TSMC was only behind by one process node and the orders started pouring in. In 10 years, TSMC caught up with IDMs (except for Intel) and the fabless semiconductor industry blossomed, enabling a whole new era of semiconductor design and manufacturing. In the last 20 years, and still today, even the remaining IDMs are being forced to go fabless or fab-lite at 28 nm and below due to high costs and daunting technical challenges.

Dr. Morris Chang in 2007.

Dr. Morris Chang turned 82 on July 10th, 2013. He is still running TSMC full time as the founding Chairman. He works from 8:30 am to 6:30 pm like most TSMC employees and says that a successful company life cycle is: rapid expansion, a period of consolidation, and maturity. The same could be said about Chang himself.

In Their Own Words: GLOBALFOUNDRIES

GLOBALFOUNDRIES is the newest pure-play foundry in the industry. In this chapter, GLOBALFOUNDRIES describes its history, mission, and future directions.

The fabless semiconductor model, first implemented in the early 1980s when IDMs figured out there was money to be made in selling excess manufacturing capacity to small chip design companies, has been an unqualified success in delivering innovation and efficiency to the electronics industry. The emergence of 'pure-play' foundries in the mid-1980s enhanced the model further still, and has enabled the success of some of the most recognized and groundbreaking names in the semiconductor industry—firms like Qualcomm, Broadcom, Marvel, Xilinx and a host of others, not to mention forward thinking product makers like Apple and Microsoft.

Indeed, the days of "Real men own fabs" seem like a distant memory in an era when a new manufacturing plant can cost more than $5 billion, process technology is approaching sub 10 nm levels, and market windows are measured in weeks not years.

As with all dynamic markets and business models, change is a constant in how ICs are fabricated. While it's fair to say that after 30 years, the foundry model has withstood the test of time, it must evolve if it is to meet the never-ending technology and economic challenges of the semiconductor industry. The fact that mobile products surpassed PCs as the largest consumer of semiconductors for the first time in 2012 underscores the macro changes that are re-shaping the landscape of electronics, and forcing a re-thinking of the supply chain. Add in seemingly inconceivable technology drivers and unfathomable price tags, and it's clear that those who can't adapt to change in the semiconductor manufacturing world are doomed.

It was against this backdrop that some visionaries dreamed of taking a new approach to the foundry model as the first decade of the 21st Century neared an end. After all, the foundry business itself hadn't really changed that much since its inception nearly 30 years prior, even if the pace of technology evolution had maintained its steady march forward, driven by the unceasing pace of Moore's Law. So the reasoning was that there needed to be some more significant enhancements of the model to better deal with the challenges at hand. The industry needed a revamp, an upgrade, a new release, and most importantly, a more global orientation—it needed Foundry 2.0. GLOBALFOUNDRIES embodied the vision when it was launched in March 2009.

Ironically, a cornerstone of the strategy centered on the very model that foundries originally disaggregated: the integrated device manufacturer (IDM). The founders of GLOBALFOUNDRIES recognized that there needed to be a tighter connection between the design process —right from the beginning at the architectural level—and the implementation in manufacturing. The 'throw it over the wall' method of Foundry 1.0 was breaking, and closer collaboration was viewed as the only way to deal with the current challenges.

So it should be no surprise that a key aspect of the GLOBALFOUNDRIES legacy can be traced back to one of the world's leading IDMs. In October 2008, AMD announced a new strategy to focus exclusively on the design phase of semiconductor product development. To achieve that strategy, AMD partnered with Advanced Technology Investment Company

(ATIC) of Abu Dhabi to create a new joint venture company designed to become the world's first truly global foundry.

2009: The Birth of Foundry 2.0

On March 4, 2009, GLOBALFOUNDRIES officially launched as a new joint venture, coupling AMD's leading-edge semiconductor manufacturing capabilities with the financial focus of ATIC, creating a new global semiconductor manufacturing foundry with approximately 3,000 employees. This formally entered GLOBALFOUNDRIES in the foundry business and armed it out of the gates with a production-proven fab campus based in Dresden, Germany—and years of seasoned experience in semiconductor design and manufacturing. AMD became its first customer.

Success with customers beyond AMD soon followed. Through the course of 2009, the company announced several new customers and new strategic partnerships, including ARM, STMicroelectronics, and Qualcomm.

In June of that same year, GLOBALFOUNDRIES took the first step in what was to become a defining element of its strategy. It was then that the company broke ground on the Fab 8 campus, the company's newest 300 mm fab in Saratoga County, New York. It was to be rightly heralded as the most advanced semiconductor manufacturing facility ever constructed.

The Chartered Acquisition

In January 2010, the company announced the completion of its merger with Chartered Semiconductor, a global semiconductor foundry company based in Singapore. At the time, Chartered consisted of about 7,000 employees, mostly based at the company's six fabs in Singapore. The addition of Chartered added more than 150 customers to the company's portfolio, afforded world class production capabilities in both mainstream and leading-edge technologies and allowed the company to offer a new platform for innovation to drive the current and future generations of semiconductor products for customers around the globe.

Overnight, GLOBALFOUNDRIES had become one of the world's top 3 foundries and the industry couldn't help but take notice. The addition of Chartered added proven experience in the workings of the foundry model, complementing the IDM legacy from the company's roots in AMD. In addition, Chartered was skilled at the partnership model and GLOBALFOUNDRIES found itself with a seat at the table of the ground-breaking Common Platform Alliance, which included IBM and Samsung in an initiative that defined new levels of collaboration among chip manufacturers and customers. The Chartered acquisition also brought with it much needed capacity and a gateway into more application areas. The Singapore operations would continue to play a major role in the company's strategy.

By 2011, GLOBALFOUNDRIES was hitting full stride, continuing to add customers and reaching significant manufacturing milestones as AMD's 32 nm processor shipments increased by more than 80% from the third quarter to the fourth quarter. In fact, GLOBALFOUNDRIES exited 2011 as the only foundry to have shipped in the hundreds of thousands of 32 nm high-K metal gate (HKMG) wafers.

New Leadership for a New Era

With the initial growing pains behind it, the company now was squarely focused on growth and implementing its vision. To that end Ajit Manocha was named CEO of the company in late 2011. A skilled leader, he brought more than 30 years of experience in the semiconductor industry, having held senior positions at Spansion, NXP and AT&T Microelectronics. Manocha was a safe pair of hands to bring the company to the next stage. Manocha understood the value of partnerships and collaboration well and quickly wove that philosophy deeper into the fabric of the company. Collaborative Device Manufacturing (CDM) became the new mantra for the model GLOBALFOUNDRIES espoused, under the name Foundry 2.0, and brand name customers and partners from around the ecosystem embraced it. In January 2014, Sanjay Jha was appointed CEO and Manocha returned to his role advising the owners of GLOBALFOUNDRIES. Jha's background was in mobile, with a long tenure at Qualcomm and a period as CEO of Motorola Mobility. Mobile is, of course, the largest market for semiconductors today and continues to grow fast.

Global Leadership at the Leading Edge

Unique to the foundry industry, GLOBALFOUNDRIES operates a global network of advanced manufacturing and technology capabilities, anchored by 300 mm and 200 mm facilities in Singapore, Germany, and the company's newest campus in Saratoga County, New York. Periodic benchmarking conducted by third parties consistently places GLOBALFOUNDRIES as a leader worldwide in the major categories for fab performance. This advanced network of manufacturing campuses and global research partnerships provides the company the ability to introduce technologies with greater process maturity than is typical of the foundry industry, enabling the fastest volume ramps in the industry.

Fab 1: Dresden

The Dresden manufacturing site is recognized throughout the industry as among the most successful leading-edge semiconductor production facilities in the world. Fab 1 represents one of the biggest international investments in Germany with a total investment to date of more than $7 billion, and about 3,000 world-class engineers, technicians, and specialists.

Fab 7: Singapore

GLOBALFOUNDRIES has two manufacturing campuses in Singapore with four 200 mm wafer fabrication plants (Fab 2, 3/5, 6) and one 300 mm wafer fabrication plant (Fab 7) located in Woodlands and another 200 mm manufacturing facility (Fab 3E) in Tampines.

The Singapore site is embarking upon a long-term strategic plan to focus on upgrading the manufacturing facilities to address fast-growing

"More Than Moore" technology areas such as MEMS, RF and analog/mixed signal with technology nodes spanning from 180 nm to 40 nm.

Fab 8: New York

For more than two decades, the focus of the semiconductor foundry industry has increasingly turned to Asia for growth and the development of new manufacturing facilities. Counter to this prevailing trend, the Fab 8 project is the first leading-edge semiconductor foundry to be built in the U.S. and one of the largest new manufacturing projects in the world. The project is a key driver in the revitalization of upstate New York's "Tech Valley" and a prime example of how advanced manufacturing can help boost the American economy. Building on a history of award-winning manufacturing facilities, GLOBALFOUNDRIES is developing the world's most advanced semiconductor wafer fab at the Luther Forest Technology Campus in upstate New York.

Less than 3 years since its formation, the company's multi-billion dollar investment in upstate New York and its extensive network of partnerships in that region began to bear fruit in 2012. In January of that year, GLOBALFOUNDRIES started running first customer silicon at Fab 8 with IBM's 32 nm SOI technology. This technology was jointly developed between GLOBALFOUNDRIES and other members of IBM's Process Development Alliance, including early stage research at the University of Albany, State University of New York's College of Nanoscale Science and Engineering.

In July 2012, GLOBALFOUNDRIES announced an extension of 90,000 square feet to the Fab 8 Module 1 cleanroom in response to strong customer demand at the 28 nm node. The extension of Fab 8 increased the cleanroom to approximately 300,000 square feet, roughly equivalent to six football fields of state-of-the-art semiconductor wafer manufacturing space. Construction work on the Fab 8 Module 1 extension project began in September and is expected to be completed in December 2013.

Technology innovation, through partnerships and extensive investment in R&D, continues to make GLOBALFOUNDRIES a force to be reckoned with. The company has established itself among the industry's elite, aggressively laying out a roadmap to 10 nm and beyond. It delivered on its promises with the announcement of the industry's first modular 14 nm offering, a breakthrough FinFET approach specifically aimed at the burgeoning opportunities in mobile application markets. This was a reflection of an acceleration of its leading-edge roadmap to give customers the performance and power benefits of three-dimensional FinFET transistors with less risk and faster time-to-market.

In addition, GLOBALFOUNDRIES began ramping its 20 nm technology in 2012, and saw significant adoption and yield improvements for its 28 nm offerings. By year end, it was clear the company no longer would take a back seat to anyone when it comes to technology.

The Emergence of a True Market Leader

By mid-2012, GLOBALFOUNDRIES had surpassed its nearest competitor and was firmly established as the world's second largest foundry in the industry rankings. The company was buoyed by the continued growth of the foundry market in general, and its unique application of the model was winning new customers at an impressive rate.

An IC Insights report in 2012 was especially significant for GLOBALFOUNDRIES as the company jumped six spots to break into the top 20 IC companies for the first time, and IC Insights projects its revenue to grow 31% over 2011, making GLOBALFOUNDRIES the fastest growing semiconductor company in the world. The firm sung the praises of GLOBALFOUNDRIES approach, noting, "It is obvious that GLOBALFOUNDRIES' current spike in revenue is being driven mostly by its success in attracting new IC foundry customers."

Focus on Collaborative Technology Development

In January 2013, GLOBALFOUNDRIES announced a new global R&D facility at its Fab 8 campus. The new Technology Development Center (TDC) Technology Development Center will play a key role in the company's strategy to develop innovative semiconductor solutions allowing customers to compete at the leading edge of technology.

The TDC will house a variety of semiconductor development and manufacturing spaces to support the transition to new technology nodes, as well as the development of innovative capabilities to deliver value to customers beyond the traditional approach of shrinking transistors. The overarching goal of the TDC is to provide a collaborative space for GLOBALFOUNDRIES to develop end-to-end solutions covering the full spectrum of silicon technology, from new interconnect and packaging technologies that enable three-dimensional (3D) stacking of chips to leading-edge photomasks for Extreme Ultraviolet (EUV) lithography and everything in between.

The TDC represents an additional investment of nearly $2 billion, increasing the total capital investment for the Fab 8 campus to approximately $8 billion. Construction of the TDC began in early 2013 and is expected to be completed in late 2014.

With the addition of Fab 8, GLOBALFOUNDRIES now operates three 300 mm wafer fabs around the world with campuses in Germany, Singapore and New York offering customers leading-edge volume manufacturing capabilities at the 32 nm and 28 nm process nodes and technology development at 20 nm, 14 nm and beyond. In addition, GLOBALFOUNDRIES also operates five 200 mm wafer fabs in Singapore, offering global customers a broad spectrum of manufacturing technology options.

Foundry 2.0 Today and for the Future

Most industry watchers have confirmed that GLOBALFOUNDRIES is a first-of-its-kind global foundry model, bringing a unique approach to leveraging assets from around the world to best meet the needs of the global marketplace. Since its inception the company has made substantial capital investments to build a truly global footprint, with manufacturing operations spanning three continents for flexible and secure supply. Today it employs more than 13,000 people worldwide, and with manufacturing centers in Germany, the United States and Singapore, GLOBALFOUNDRIES is delivering advanced technologies to market in high volume and mature yield faster than any other foundry in the world.

This global manufacturing footprint is supported by major facilities for research, development, and design enablement located across the U.S., Europe, and Asia, with offices in Abu Dhabi and corporate offices in Silicon Valley. The collective strength of these operations is unprecedented for a semiconductor foundry and unparalleled in the industry.

A final, perhaps symbolic, milestone was reached in 2013 when AMD completed its divesture of the remaining 14% stake it had in GLOBALFOUNDRIES. The transition from IDM to foundry was now complete. Just four short years from its founding, GLOBALFOUNDRIES is wholly owned by ATIC and is firmly entrenched as the world's second largest independent semiconductor foundry that has written an entirely new chapter in the history of an industry.

Chapter 6: Electronic Design Automation

There is no doubt that EDA has been a key enabler of the fabless semiconductor industry. EDA and IP are not so much the tail that wags the dog, rather they are like the heart of an elephant, tiny in comparison but without which there is no elephant. In this chapter, we lay out (manually, as it turns out), the history of EDA as we see it, taking you through the five historical phases of semiconductor design.

EDA, Phase One

From the earliest days of integrated circuits until the mid-1970s, chips were designed manually, with no automation of any part of the design, layout, verification, or mask preparation. The masks used for photolithography were actually hand cut with X-ACTO knives out of a self-adhesive red plastic film called Rubylith. Rubylith is no longer used in IC manufacture, but it is still used to make masks in many other areas of graphic design. When it was still used in IC design, the Rubylith was stuck onto transparent paper or plastic in patterns that defined the transistors and interconnect and then photoreduced to get the actual masks. As chips got larger, this process became more and more unwieldy, both because of the number of pieces of Rubylith required and because of the sheer size of the sheets onto which it was stuck. It became clear that some method of automation would be required very soon.

About this time, in the mid-1970s, three companies entered the market for IC layout automation—Calma, Applicon, and ComputerVision. We consider this the first phase of the EDA industry. Their products replaced manual design layout with a computer system that allowed layout engineers to trace out the mask shapes on the screen. When the design was released to manufacturing, the software would save the layout to a magnetic tape that a photoplotter could use to create the actual mask. This process was called tape-out. Even though tapes are obsolete, and photoplotters have been replaced with e-beam mask-making machines, releasing a design to manufacturing is still called tape-out. Tape-out is the culmination of months or even years of work. It is the point-of-no-return; a multi-million dollar wager that every "T" has been crossed and every "I" dotted.

The complexity of chips has long demanded the use of computer software to simulate chip behavior, design the physical layout of the chip, verify the functionality, and ensure the design can be manufactured. The broad and enormously sophisticated software programs that enable the creation of all chips fall into the category of EDA software.

The original EDA companies (Calma, Applicon, and ComputerVision) sold the hardware for the graphical layout with the software bundled in. Because of this, the EDA industry business model was based on a hardware business model—the customer purchased the hardware and paid an annual maintenance charge to keep it running. This remained the model for the EDA industry for many years, even after it became a pure software business. There was even, for a time, a worry that the falling price of hardware would pull the price of software down too, because software was often bundled with hardware. In the 1970s, unlike today, software was cheaper than hardware. As the relative value of hardware and software inverted, the EDA industry worried that their customers would not pay more for software than for the hardware on which it ran. Today this would strike people as comical because computer hardware is a commodity and EDA software (and other enterprise software like databases) costs hundreds of times more.

Until about 1980, semiconductor design was only done inside semiconductor companies. They decided what to make, then systems companies could buy those chips and design products around them. Apart from layout tools and some circuit simulation, most other software design tools were largely developed in-house by internal computer-aided design (CAD) groups.

For example, Hewlett-Packard created its own "integrated graphics system" called HP-IGS. Long-time industry insider, Randy Smith, who's first EDA job was working on the HP-IGS, says the application software ran on an HP3000 computer (really a 6-foot tall business computer) while the graphics were processed and displayed on an HP1000 microcomputer.

Increasing the level of automation was seen to be a competitive advantage and a company's internal software was considered part of their "secret sauce." But the world was about to change, and the initial catalyst for that change was largely a single, very influential, book.

EDA, Phase Two

In 1980, Carver Mead at Caltech and Lynn Conway at Xerox Palo Alto Research Center (PARC) published their *Introduction to VLSI Systems*. This book marked the first time that the details of designing an integrated circuit were openly available to people outside of the semiconductor companies themselves. Universities, research centers, and system companies could suddenly consider designing their own integrated circuits rather than buying chips from the semiconductor companies. Although chips at the time could contain about 5,000 gates, which was too small to make most interesting systems, the basic idea of Moore's Law, that the number of gates would double every couple of years, was by then well understood and its implications were becoming clear. Semiconductor technology would grow, and it would affect our lives in ways that were open to discovery. Any electronic system could eventually be implemented with just a few chips, and very inexpensively.

After the publication of *Introduction to VLSI Systems*, computer scientists flocked to integrated circuit design. Unlike the VLSI designers in the semiconductor companies, who often had a deep understanding of electrical engineering and semiconductor process, the computer scientists did what came naturally: they created simplified abstractions, specifically hierarchy, to manage the complexity of chip design. The old ways of designing, even with computer-aided layout on systems like Calma, could not keep up with the growing complexity of designs with thousands of gates, let alone the tens or hundreds of thousands that was anticipated based on the Moore's Law projections.

These abstractions created by the influx of computer scientists drove one of the most important evolutions for the semiconductor industry—the creation of ASICs, the chips designed for a specific use, as opposed to the all-purpose chips that the semiconductor companies were creating at the time.

As we mentioned in the ASIC chapter, the ASIC companies initially produced many of their own design tools, but as the design processes became more standardized, a more generic ASIC methodology emerged. The ASIC flow essentially split the process into two distinct parts—front-end and back-end. The front-end design was, and is still, done by the systems companies and included the chip architecture and simulation. The system company would select a library of standard cells for the design and specify how to hook them up by using a special graphical software tool. (Standard cells are a group of transistors that perform a logic or storage function, implemented as fixed-height elements.)

The systems companies used simulation software to ensure that the design performed as intended. This combination of library cells and how they were interconnected, called the netlist, was then shipped to the ASIC company who would do the physical design of the chip, also known as back-end design.

Because the initial ASIC methodology split the design into two parts—the front-end done at the systems company and the back-end done at the semiconductor company—two distinct types of EDA companies flourished, and is the industry we still have today.

So at this point, the industry had a design flow that was bifurcated into front-end and back-end, and chips that continued to grow in complexity. Both of these factors led to increased specialization of all aspects of the ASIC flow and opened the door for a new crop of EDA companies—Daisy Systems, Mentor Graphics, and Valid Logic Systems.

All three created tools to handle the front-end design tasks, mainly schematic editing and simulation. Daisy and Valid continued the existing Calma business model of selling hardware on which their software ran. Mentor Graphics however, created software that ran on workstations made by the company Apollo (who would be purchased by Hewlett-Packard in 1989). In the early days, all three provided roughly equivalent software for schematic capture and for verifying designs through simulation.

The back-end design started with the netlist, timing information, and process information. The placement of standard cells and the signal routing between them was accomplished with the help of software developed in-house by some ASIC companies or by one of the EDA companies—such as Silvar-Lisco and Tangent Systems—that specialized in the hard-to-master placement and routing algorithms.

None of the original big three layout companies—Calma, Applicon, or ComputerVision—made the transition to this new world of ASIC in which schematic design and automated place and route, as opposed to manual layout, became the key enabling technologies. They were each eventually absorbed into other companies and all the technologies they developed, except for one, fell into the trash bin of history.

The one technology that did not vanish was Calma's, called simply the Graphic Design System or GDS, which they originally released in 1971. The second version of it, introduced in 1978, was thus called GDSII. Back in those days, computer hard drives couldn't hold all the designs in progress, nor was there good networking between systems, so design data was kept on a magnetic tape in the GDSII stream format. This was essentially a daily back up format. The layout designer would work on the design, then she (many layout designers were women) would save it to the disk of a free system. This GDSII format became the de facto standard for moving design data around between systems. Amazingly, almost 40 years later, that format is only just starting to be superseded (by a new standard called OASIS) as the standard format for moving design layout between tools or between design house and mask shop.

By late 1980s, a lot of semiconductor design was done using the ASIC methodology, with the system companies doing the front-end and the semiconductor companies doing the back-end physical design. This was the second phase of the EDA industry.

EDA, Phase Three

The third phase of the EDA industry was driven by two factors. Firstly, it became unfeasible for every semiconductor company to develop all their own place and route tools internally, so more and more companies discontinued their internal development in favor of buying EDA tools from external EDA companies. Secondly, it became possible for system companies to do their own back-end design in addition to the front-end, so they didn't rely as heavily on ASIC companies to complete the lphysical layout of their designs.

These two trends meant that the EDA industry no longer focused just on the front-end where Daisy, Mentor and Valid were strong, but began to focus on products for the more complex physical design.

The most important company of this third phase of the EDA industry was Cadence Design Systems. In fact, Cadence was created as the merger of two earlier companies, SDA and ECAD, who produced tools for physical layout and tools for verifying that they were laid out correctly. Cadence became the dominant company for custom design, and following its acquisition of Tangent Systems in 1989, also for place and route.

Just as Calma, Applicon, and ComputerVision vanished at the start of the second phase of the EDA industry, Daisy, Mentor, and Valid were overtaken by the new EDA companies like Cadence. Daisy made an unwise merger with Cadnetix Corp in 1988 and soon after went out of business. As an interesting side note, the defunct Daisy was picked up in 1990 by Intergraph Corp, which had been the parent corporation of Tangent before Tangent was acquired by Cadence the year before. Cadence also acquired Valid in 1992. Intergraph used Daisy/Cadnetix technology in their new subsidiary called VeriBest, who was acquired by Mentor in 1999. Got all that? Of the three, only Mentor survived intact, but it spent years re-architecting itself for the new era and only in recent years has it acquired a full portfolio of physical design and verification tools.

There were two other companies that were significant during this third phase of the industry—Gateway Design and Arcsys. Gateway created the simulation language called Verilog and produced a high-performance simulator. Cadence acquired them in 1989 for $72 million, which at the time, seemed an enormous price. In hindsight, it is one of the most successful EDA acquisitions ever. Verilog turned out to be extremely important in the fourth phase of the EDA industry.

Arcsys was created to compete with Cadence in automatic place and route, which was then the biggest and richest sub-segment of the EDA industry. Arcsys acquired a physical verification company called Integrated Silicon Systems (ISS) in 1995 and changed its name to Avant! (pronounced "ah-VAHN-tee"). They became the second company behind Cadence in the physical design area.

Arcsys/Avant! became infamous for another reason: its first product was built on source code for the underlying database that was stolen from Cadence. This led to FBI raids and years of litigation before both the criminal and civil processes eventually concluded with jail terms and hundreds of millions of dollars in restitution and fines.

EDA, Phase Four

So the third phase of the EDA industry saw semiconductor companies replacing their internal tools by, largely, Cadence and Avant! software. The fourth phase of the EDA industry was the transition to synthesis-based design, which only became mainstream in the mid-1990s. Graphical schematic-based design was replaced, except for analog and some other specialized areas, by language-based synthesis with astounding leaps in design productivity. Synopsys won this part of the market but there were many competing technologies in the beginning: SILC, Autologic, Trimeter, and others. Much later, Cadence tried to develop its own synthesis product, called Synergy, but it was never successful.

Synopsys built their logic synthesis business in stages. First they created logic optimization products that took the netlists from graphical schematic capture and improved them. Then they produced a tool that would read Verilog, automatically create a netlist that had the same functionality, then optimize it. This synthesis-based approach to design is still the mainstream today, and the synthesis tools have improved in performance, capacity and other dimensions.

Around 1998, at the end of this fourth phase of the EDA industry, the landscape was as follows: in custom design Cadence was dominant; in place and route the market was split between Cadence and Avant!; in simulation Mentor, Cadence, and Synopsys had products, but Synopsys was dominant; in physical verification Cadence led, but Avant! and Mentor both had products too.

EDA, Phase Five

The fifth phase of the EDA industry, which brings us up to the present time, is the era of full-service EDA companies. Through the mid-1990s, most semiconductor companies built their design flow with point tools for each stage of the design flow, regardless of how many EDA vendors they had to buy from. This strategy grew out of necessity because no one EDA company had all the tools they needed. Synopsys and Mentor had no place and route, and Cadence had no synthesis, for example. It was impossible to put a whole flow together from one EDA company.

In pursuit of full-flow software offerings, Cadence acquired synthesis technology from the companies Ambit in 1998 and Get2Chip in 2003. Synopsys acquired Avant! in 2001, thus concluding Avant!'s legal woes. Mentor was slower to round out its offerings by finding place and route tools, but it made up for it when, almost overnight, its Calibre physical verification software replaced Cadence's product (Dracula) as the industry standard.

As EDA companies built portfolios of software to cover the entire design flow, semiconductor companies switched from putting together their own flows with best-in-class point tools from a mixture of companies, to picking a single primary supplier, typically Cadence or Synopsys, and supplementing with some additional tools such as Mentor's Calibre. Business deals went from selling individual licenses to selling a huge bundle of capability, sometimes called "all-you-can-eat." Not everyone transitioned smoothly to new licensing terms. During this time, Synopsys overtook Cadence to take the #1 position in the EDA market.

During this period, in 1997, a new company called Magma Design Automation was launched. Magma pursued new algorithms that allowed for the merging of synthesis and physical design, selling a fully-integrated "physical synthesis" product from day one. Magma used over $100 million in venture capital before going public in 2001 and did get some traction against the incumbent EDA companies. They broadened their product portfolio, adding internally-developed circuit simulation and custom design products, but they never managed to approach the size of the bigger companies. However, after a few business-related missteps, Magma's revenues suffered and they were acquired by Synopsys in 2012 for $507 million. A complete list of acquisitions by EDA companies is available on SemiWiki.com.

Throughout the history of EDA, a lot of the innovation has happened in small venture-funded startup companies. Acquisition has always been a sound exit strategy for EDA startups. Usually, the bigger EDA companies wait until their smaller competitor's technology is proven through market acceptance, and then they acquire them. The big EDA companies, to different degrees, were unable to develop completely new products, and found it even harder to get into the channel with new products because the switch to big bundled deals didn't leave space for a couple of licenses of an immature product. The EDA industry is largely one of spin-offs, startups, and acquisition. That makes it full of innovation, and of intrigue. There could be an entire new industry dedicated to predicting who will buy whom.

Between 2008 and 2012, venture funding for EDA startups dropped dramatically. One source (from information tracked by Mentor Graphics) states that venture capital going into EDA went from $169 million in 2007 to only $29 million in 2010. Funding has gone disproportionally to social media, which can offer higher returns. Investors also shy away from the sticky technical problems for which modern chip design is the poster child. The technical hurdles faced by EDA startups are large, as are costs of developing them and especially of bringing them to market. Still, there is evidence that VC funding is returning to EDA; there are more EDA startups than fabless startups. One reason is that while the returns are historically modest for EDA investments, the capital costs for an EDA startup are also low. High performance computers are now available on your lap and coffee shops have replaced office space.

EDA startups face great challenges trying to create solutions to problems—like designing 3D chips and doing double or triple patterning—that are simply too big to be done by a small company. These technical challenges require changes to dozens of tools throughout the flow and cannot be solved by a single point tool dropped into a pre-existing flow. As a result, a lot of innovation is now taking place in the big EDA companies and is pushed out to the customer by the simple fact that each new process node requires significant changes to all the design tools. You can't, for example, get through a 20 nm design using 28 nm design tools.

At the beginning of 2013, the EDA industry has three dominant players and a robust supporting cast of dozens of smaller startup companies, perhaps even a couple of hundred depending on how you count. There are also three medium sized EDA companies, Atrenta, Apache (which is a subsidiary of a much bigger company ANSYS), and Silvaco. You can read their histories on SemiWiki.com.

The following chapters are written by the big three—Synopsys, Cadence, and Mentor. We asked them to tell their histories in their own words, including their creation and how they fit into the EDA industry.

In Their Own Words: Mentor Graphics

> Mentor Graphics is the oldest EDA company still in operation. They've seen, and helped shape, changes to technology and business models for over 30 years. In this section, Mentor shares their history, technology, and their role in the developing the current EDA business environment.

In 1981, Pac-Man was sweeping the nation, the first space shuttle launched, and a small group of engineers in Oregon started not only a new company (Mentor Graphics), but also, along with a handful of other companies, helped launch an entirely new industry, EDA.

Mentor Graphics founders—Tom Bruggere, Gerry Langeler, and Dave Moffenbeier—left their comfortable, secure jobs at Tektronix, Oregon's largest electronics manufacturing company at the time. They were all bright, ambitious, 30-somethings determined to take advantage of the nascent area of computer graphics.

They quickly zeroed in on the promising market of computer-aided engineering (CAE): the automation of schematic capture and simulation for engineers designing complex electronics systems including printed circuit boards. The founders spent their first months traveling exhaustively, interviewing numerous high tech companies about their design challenges. A preliminary idea for a CAE product started to emerge during this process. Eventually they decided to focus solely on developing CAE software and use commercially-available workstations for the hardware.

Other EDA startups at that time hewed to the time-honored business model of creating a vertically integrated solution, designing both hardware and software, stretching precious resources across the two domains.

This decision was risky for many reasons, more so because Mentor's founders chose the Apollo workstation as their hardware platform while it was still only a specification. They personally knew the Apollo founders and trusted the company could create, on schedule, a new type of computer that combined the time-sharing capabilities of a mainframe with the processing power of a dedicated minicomputer. Their calculated gamble paid off. Creating software from scratch that met specific customer requirements, while using commercial hardware, proved to be a key advantage over other fledgling CAE competitors in the early years.

Mentor's Founders. Clockwise from top left: Charlie Sorgie, Dave Moffenbeier, Steve Swerling, Tom Bruggere, Jack Bennett. Seated from left: Gerry Langeler, Rick Samco, Ken Willett, John Stedman.

The Apollo computers were delivered in the fall of 1981 and the Mentor engineers began developing their software. Their goal was to unveil the first interactive simulation product, IDEA 1000, at the Design Automation Conference in Las Vegas the following summer. Rather than being lost in the crowd on the show floor, they rented a hotel suite and invited participants to private demonstrations. Actually, invitations were slipped under all the hotel room doors at Caesar's Palace. Invitations were passed out indiscriminately to vacationers and conference-goers alike because Mentor didn't know which rooms were DAC attendees. The demos were very well received (by conference goers, anyway).

One of the co-founders, Gerry Langeler, vividly remembers the response to those first demos:

> "I made the presentation while one of our engineers worked the keyboard of the CAE workstation, and the demonstration software performed flawlessly. I watched faces go from casual interest to intense scrutiny and on to slack-jawed disbelief and undisguised enthusiasm. I saw prospects turn into customers. Word spread. People crowded into the room. People stood in the hallway craning their necks to catch a glimpse of our demonstrations. Over the course of the conference, perhaps as many as half the delegates found their way to our suite. And then came the jackpot: a purchase order for one system was delivered to us in our suite. We were bona fide. People had bought the something we had built."

IDEA 1000 was quickly extended to include a suite of capabilities that were enthusiastically adopted by engineers creating complex designs for silicon ICs or printed circuit boards (PCBs). The emergence of these tools corresponded to the rise of ASIC design. These highly crafted digital ASIC chips required extensive verification, to ensure correct operation in their end-system environments. For example, IDEA Station provided complete, automated schematic capture. Then with Mentor's QuickSim analysis for gate-level simulation, designers were able to examine the circuit functionality before committing to a physical prototype, enabling them to quickly iterate and improve the quality of the design. IC Station then was used for the place and route of custom IC designs, reducing the task from weeks or months to just hours. And Board Station performed a similar place and route functionality for PCB designs.

All three of these products—IDEA Station, IC Station, and Board Station—became market leading EDA products. Companies across the electronics industry became major customers, eager to tap the powerful design, analysis, and implementation capabilities of Mentor's offering. Customers included numerous computer companies from Apollo computer to NEC; semiconductor companies such as Motorola and Texas Instruments; consumer and telecommunications companies such as AT&T, Canon and GM/Delco; and large aerospace companies including Boeing, Rockwell and Lockheed.

With this strong industry response, Mentor became the fastest U.S. startup company in history to reach $200 million in revenue, reported its first profit in 1984 and went public the same year. To service all these customers around the world, Mentor Graphics began opening offices across the United States, and in Europe and Asia. Throughout the 1980s, Mentor grew and became one of the most profitable and largest U.S. startups in the 1980s. Mentor crested the $400 million mark in 1990 and seemed poised for continued success.

Painful but Healthy Realignments

Unfortunately, all this rapid success for Mentor had an unintended, and potentially threatening, side effect. The company became a victim of what has since come to be known as "The Innovator's Dilemma" (from Clayton Christensen), listening to their customers' desire for a single integrated interface design environment from start to finish. This pursuit of a "complete solution" flew in the face of what sophisticated customers clearly wanted—point tools that were best-of-breed which could easily be integrated into existing flows to meet each emerging design need. This became increasingly important as EDA startups accelerated their introduction of innovative design capabilities that needed to be integrated into a design flow.

In addition, the change in the EDA model from bundling software with hardware changed. Mentor made the tough decision in 1991 to stop bundling Apollo workstations with the software and to support other hardware, such as Sun. Mentor revenue peaked at $435 million in 1990 and then fell as the hardware business disappeared.

Meanwhile, the integrated "framework" under development and now known as the "Falcon Framework (also called Version 8.0)" proved to be an overly ambitious undertaking. While Cadence and Mentor pursued similar approaches, Mentor bet 100% on success of the "framework" approach and provided no backup for non-framework customers to build their own environments out of "point tools." The disruption associated with development difficulties led to lots of changes, including the recruiting of a new CEO. Wally Rhines, who was Executive VP in charge of Texas Instrument's $5 billion semiconductor business, surprisingly decided that Mentor offered an opportunity for innovation and growth, heralding the beginning of a new direction away from the "framework" approach.

Back on Track

Rhines came to Mentor with an extensive understanding of the electronic design process, having managed most types of semiconductor businesses, as well as a $1 billion minicomputer and computer peripherals business. His success at TI had been propelled by the success of digital signal processing where he first supervised the development of a chip set used in "Speak 'n Spell," a talking educational product, and then the conception, development and commercialization of a complete family of digital signal processors, the TMS 320 family, which eventually evolved to become nearly half of TI's total revenue.

Rhines spent his initial time at Mentor stemming the bleeding from Version 8.0 and redirecting the company to a strategy that could easily accommodate non-framework based point tool developments, both by Mentor and by third parties, into useful design solutions. One month after his arrival, Mentor acquired Checklogic, which eventually evolved to the industry's leading design-for-test solution and stimulated two of the most significant discontinuities in DFT—compression and cell-aware test. Three months after his arrival, a major initiative to develop a new generation of physical design verification was kicked off, since Mentor's Checkmate product, which had achieved only moderate success against Cadence Dracula, had been licensed from Wally himself when he was at TI (after Cadence acquired ECAD which provided the Mentor Dracula OEM product), and Mentor had the capability to develop its own unique approach to physical verification.

These needs quickly evolved into a new strategy for Mentor, completely different from the "own the whole flow" strategy of Version 8.0:

1) Focus where you can be #1, but support open standards so that you can be easily integrated into all design flows
2) Look for design discontinuities to replace existing solutions
3) Identify new emerging problems and develop the tools that will be needed before the problems become big issues

While Mentor had lost time, momentum, and its #1 market share position to the Version 8.0 diversion, the company quickly began making up for lost time. Having led the EDA industry in gate-level simulation with Quicksim, Mentor developed an early RTL simulator for the newly emerging multi-company VHDL standard which was supported by much of the customer base as a counter to Verilog which was developed as a proprietary simulator. With the acquisition of Model Technology, Mentor was able to provide the industry's first direct-compile simulator which, because of its "single kernel," quickly supported VHDL, Verilog and subsequently System Verilog, C++, System C and other languages. Mentor was therefore able to sustain its #2 market share position in RTL simulation for the next 18 years (except for the years that it became #1, according to GSEDA).

Meanwhile, the Calibre product team of three core people, Laurence Grodd, Koby Kresh, and Robert Todd, built upon their long experience supporting Dracula and Checkmate to develop a totally new approach to physical verification using "hierarchy" as a way to dramatically impact performance. Operating as a virtual "skunkworks" in Mentor Graphics, the team worked with customers that were outside the targeted IC Station user base, running benchmarks without Mentor management awareness. By late 1996, word got out that Mentor had something really unique and the adoption of Calibre took off as users of existing physical verification software found that they couldn't verify large designs at 250 nm and below. A long series of innovations followed over the next several years, leading to a total of 48 patents granted to the Calibre core team and eventually Calibre expanded into a full platform for physical verification, analysis and design-for-manufacturing. The team made sure that Calibre was so well integrated with competing design flows that compatibility hardly ever became an issue.

Other Changes in Leadership and Direction

In 1996, Mentor recruited Greg Hinckley, former Senior VP and CFO of VLSI Technologies to become COO (and CFO) at Mentor. Greg's "out of the box" thinking, analytical skills, and business experience made him and Rhines close partners in the management of Mentor and accelerated the focus on innovation. From his experience at VLSI, Greg attracted Don Maulsby to manage Mentor Worldwide Sales and Henry Potts to run the PCB business. While most companies believed that PCB was a "dead" EDA business, Greg and Wally combined their contrarian views and placed new emphasis on emerging opportunities in system design. Building upon a #1 market share position in PCB, Henry Potts expanded the system design business into totally new areas like signal integrity, thermal analysis and, most importantly, transportation system design. Complemented by Serge Leef's interest and developments in automotive network analysis and an emerging new standard called AUTOSAR, the systems design activity thrived and became the fastest growing major business during the 2000s.

To complement hardware design, Mentor entered the embedded software business via acquisition of Microtec (the largest embedded software company and owner of the VRTX RTOS) in 1996 and later companies like Accelerated Technologies (owner of the most widely used RTOS, by number, NUCLEUS).

By this point, Mentor's strategy of building upon its #1 market share positions, like PCB, and identifying technical discontinuities in existing flows such as Calibre, was producing results and the revenue in the late 1990s and beyond grew somewhat faster than the overall industry. But it was the identification and support for newly emerging design problems—the third leg of the strategy—that fueled the next wave of growth. These included:

- Emergence of the need for emulation for both hardware and software verification
- Adoption of embedded software development environments by chip and system development teams

- Application of the basic tools of electronic design automation to system design, particularly for planes, trains, cars, and distributed networks
- High-level ESL design
- Resolution enhancement for design for manufacturing (DFM)
- Adoption of open source software and LINUX-based embedded development environments
- Enhancements to existing capabilities such as: Compression and cell-aware techniques for DFT; Push-button formal methods for simulation; Intelligent test benches; High-level power analysis, and many more

One of the most exciting developments was the invention of compressed test by Janusz Rajski. The automatic test equipment industry (ATE) had been unable to keep test cost in line with the overall manufacturing cost per transistor, making test a rapidly increasing portion of the total cost of manufacturing. Janusz developed a technology that would initially allow first for 10X compression of test patterns, and similar reduction in test time, and then improved it to more than 200X on average, with a clear roadmap to 1000X. This was a major factor in Mentor becoming the #1 provider of EDA test solutions. More recently, development of Cell-Aware ATPG is proving to be a similar game changer, offering order of magnitude improvement in test quality as well as the unique ability to reliably test parts that contain FinFETs.

Open Standards Pave the Way

A key enabler contributed to Mentor's role as the only #1 EDA company to fall from grace and then recover to become a top contender once again. That was the role of standards. Because of the failure of Version 8.0, Mentor became an aggressive supporter of open standards to integrate tools and platforms into flows, especially flows dominated by competitors. Mentor made this part of the accepted culture, giving as well as receiving. Whenever a Mentor approach became popular, it became a candidate to donate to a standards organization. Among the list of contributions that were wholly, or partly, from Mentor are: UPF, SystemC, UCIS, OASIS, JEDEC, IJTAG, VHDL, OpenDFM, and

OpenPDK among others. This preoccupation with standards led to interesting competitive situations.

For example, TI, Nokia, and Mentor developed a standard for power management that Mentor implemented in its simulators about year 2000. As other customers found need for it, Mentor enlisted Synopsys and Magma to join Mentor in an Accellera-sponsored standardization. Similarly, when Mentor's Advanced Verification Methodology (AVM) approach to simulation gained popularity versus a proprietary approach called Verification Methodology Manual (VMM), Mentor and Cadence joined forces to provide a design methodology called Open Verification Methodology (OVM). A widely-accepted standard was born and became the basis for what is now called Universal Verification Methodology (UVM). One of the most interesting was the early adoption of System Verilog. Synopsys solicited Mentor's participation, believing that Mentor would be relatively passive. When Mentor became first to market with System Verilog by more than a year, the industry changed and so did Mentor.

Litigation Strikes

Mentor avoided litigation wherever possible, but sometimes it just wasn't possible. One case occurred in 1997. Mentor was an early pioneer in the technology of hardware acceleration, or emulation, introducing a key product in 1988. Subsequently, Mentor sold this technology to a startup called Quickturn. This proved to be an expensive mistake. Rhines, who had been a big fan of emulation since his integrated design experience starting in the mid-1970s, initiated emulation development as soon as he arrived at Mentor. As Mentor became successful, it had to defend itself against its own patents that had been sold to Quickturn (now Cadence). Subsequently, departure of the emulation design team led to other patent litigation, all of which detracted from more productive uses of time. But the story has a happy ending, as Mentor's engineering team developed the industry-leading Veloce hardware acceleration platform and reclaimed the number one position in emulation in 2013.

System Design Focus

Because EDA in the early 1980s included both chip and board design, Mentor won the race very early for the hearts and minds of the automotive and military/aerospace customers. These customers developed stable, broad infrastructures for designs that didn't change quickly. Recognizing that strength, Mentor paid a lot of attention to these customers and actually invested a higher percentage of R&D in the last 15 years in system design than in EDA as a whole. The result was a firm foundation when these system companies began automating their design processes in the same way that semiconductor companies had done several decades before.

For Mentor, this included tools to automate virtual design of cars, trains, planes, and bigger systems. Introduction of design tools for automotive interconnect design in 1992 evolved to a major thrust under Martin O'Brien in year 2000 and the evolution of the Capital family of enterprise design tools, from architectural concept through electrical design/analysis, cost tradeoff analysis, manufacturing set-up/bill of materials, and service and support. This was complemented by families of automotive design products developed by Serge Leef's group that included network analysis and the industry's first AUTOSAR design tools to support an emerging automotive design standard. The addition of an open source software team under Mark Mitchell and an embedded software development capability under Scot Morrison provided Mentor with unique capabilities for a rapidly growing transportation market. Today, the application of EDA to systems design is driving growth that is substantially greater than the growth of adoption of traditional chip design automation.

Innovations in Higher-Levels of Abstraction

Mentor was a very early investor in the emerging Electronic System Level design abstraction, referred to as ESL. For about 15 years, Mentor was the only major EDA company with substantial revenue in this product space. Development of Seamless, the EDA industry's first successful hardware/software co-verification product, was an early achievement, as was Catapult C, a high-level synthesis product. But it was evident that a more complete solution to the high-level design challenge was needed. One of Mentor's competitors began bidding a high price for Calypto, a company that had developed high-level power analysis and optimization tools and it was clear that an integrated flow with Calypto would have value for customers. Because Mentor couldn't offer such a high price, it proposed an in-kind approach, spinning off the Catapult business into a combined entity with Calypto and retaining ownership in proportion to the revenue and profit contribution. The combined company flourished and operated independently of Mentor. Meanwhile, Mentor retained Vista, a high-level design product developed by its Israeli team under Guy Moshe, and found itself early in the race to high-level power/performance analysis.

Doing What Others Don't Do

With the arrival of Greg Hinckley at Mentor, Wally had reinforcement for his natural contrarian inclinations and Mentor increasingly explored areas outside the traditional space of EDA. One of the early efforts involved embedded software, beginning with the acquisition of Microtec in 1996 but continuing with increasing emphasis as embedded software became a more important part of every electronic design team. Thermal analysis and computational fluid dynamics was another area that attracted Mentor interest, becoming more than 5% of revenue in 2013. And Mentor fearlessly entered totally new areas like hardware for analysis of thermal inertia and for designing lighting systems. The continuing emphasis on systems design opens the door to dozens of new possibilities in the future.

Single-Vendor Flows

While Mentor championed best-in-class tools and the capability to smoothly integrate tools from third parties into Mentor flows, there was at least one case of customer support for a single-vendor approach. This was in the area of printed circuit board design and manufacturing. As PCB design matured, and Mentor's market share approached 50%, many customers began pushing for a flow that would integrate everything from concept through manufacturing deployment and yield improvement. The acquisition of Valor in 2010 completed that flow and led to a variety of new capabilities.

| Walden Rhines | Gerry Langeler | Tom Bruggere | Dave Moffenbeier | Gregory Hinckley |
| Chairman & CEO | Original Founder | Original Founder | Original Founder | President |

Mentor's founders with the current company leaders in 2001 during Mentor's 20th anniversary celebration. Photo courtesy of the Wilsonville Spokesman.

Calibre Goes Beyond Verification

As Calibre became a de facto standard for physical verification, more and more capabilities were added. Design-for-manufacturing became a big issue and a wide variety of tools for modeling lithography hot spots and yield limitations were required. Optical proximity correction continued to evolve to new levels of sophistication.

But one of the most interesting was the evolution of yield enhancement capabilities. A number of customers suggested that because Mentor was the leader in both design-for-test and physical verification, we should consider combining the two databases to look for systematic layout problems. That is, take the massive volume of test data and correlate it with physical layouts to look for "outliers" that showed statistically anomalous failure rates. The family of yield enhancement products introduced in 2005 grew to be a cornerstone of yield enhancement (and profit improvement) for the semiconductor manufacturing industry.

The Future

What makes the outlook so bright for Mentor Graphics? After all, the semiconductor industry is only growing at a 5-7% rate and the EDA industry has traditionally grown at the same rate as semiconductor R&D, being limited to about 2% of total semiconductor revenue.

There are two basic phenomena that provide for excitement and future growth. First, the semiconductor industry technology regularly adopts new technologies, each of which leads to a spurt of growth for the newly needed EDA tools. Recent examples of this have been:

- The need for reliability analysis tools for electrostatic discharge and electro migration analysis that made Calibre PERC a de facto approach.
- Evolution of 3D IC production that required a new set of verification tools (Calibre 3D) and a new approach to design-for-test with the Tessent suite of products.
- New requirements for quality that led to the development of Cell-Aware ATPG to detect transistor-level defects with gate level test patterns, and many more.

The second phenomenon is the inevitable adoption of EDA technology by the systems industry. To a first approximation, the systems industry uses EDA at about the same level as the semiconductor industry did in the 1960s. Building and testing physical prototypes has been the standard way that industries begin their path to design automation. Electronics sold with systems like cars, planes, and industrial equipment total around two trillion dollars per year, compared to the $300 billion of semiconductor electronics sold each year. As the systems industries adopt EDA, it's likely that they too will spend a percentage of their revenue in the quest for automation, making today's EDA revenue seem trivial by comparison. Mentor's early history in system design, as well as its survival as the oldest major EDA company, provides the basis for leading this next revolution.

What's Next in EDA?

The history of EDA industry growth has been driven by the emergence of new design challenges. The early generation of schematic capture and simulation was quickly augmented by PCB design, IC place and route ,and physical verification. In the last ten years, virtually all EDA industry growth has come from totally new design methodology requirements, e.g. sale of IP blocks, resolution enhancement, ESL, formal verification, design for manufacturing, and a few more. The best assumption is that the future will evolve as it has in the past, i.e. solutions to new design problems as well as the application of EDA technology to challenges in other areas of design.

With the evolution of IC design into the 14, 10 and 7 nm realms, there will be requirements for analysis of new physical design problems. Examples include reliability, electromigration, thermal effects, stress, EUV resolution enhancement, and yield analysis. Even larger will be the adoption of electronic design automation by system design companies that have been able to get by with semi-manual methods in the past. Automotive and aerospace applications are the most obvious since the electronic complexity of cars and aircraft is increasing so rapidly, probably 5% per year or more. How long will it be before we simulate the electronic behavior of an entire car or plane? A long time. But the capability to design and optimize the electrical interconnect, verify correct operation of safety, environmental and security features, manage the tradeoffs in cost and weight versus performance and provide a complete electronic data base that can be used by automotive engineering, manufacturing and service is already here and will be a big part of EDA industry growth in the next ten years.

In Their Own Words: Cadence Design Systems

> Cadence Design Systems has been a key player in the semiconductor and electronics ecosystem for a quarter century. In this section, Cadence shares their history, technology, and their role in developing the EDA business environment.

Cadence is a leading EDA supplier with comprehensive solutions for custom/analog IC design, digital IC design, functional verification, and IC packaging and printed circuit board (PCB) design. In addition to these "traditional" EDA domains, Cadence is also developing new solutions for system-level design and verification and is adding to a growing portfolio of design IP and verification IP. Cadence today has deep partnerships with customers, foundries, and IP providers.

In 2013, Cadence® celebrated its 25th anniversary. That's because two mid-sized EDA vendors—SDA Systems and ECAD—merged in 1988 to form Cadence. However, the Cadence story goes back well before 1988, to the founding of ECAD and SDA Systems in 1982 and 1983, respectively.

ECAD was founded by Glen Antle and Paul Huang. Both were working at the Systems Engineering Lab (SEL) when the CAD group developed a new, and very fast, algorithm for design rule checking (DRC). Huang directed the development of this IC physical verification technology. In 1982, Gould Inc. bought SEL, and Gould granted Antle and Huang the marketing rights to the technology.

Antle and Huang launched ECAD, and the DRC software became Dracula, one of the EDA industry's best-known products in the 1980s and 1990s. ECAD also developed Symbad, an IC layout product line.

ECAD was an unusual EDA company in the 1980s because it sold only software that supported workstations and computers from multiple providers. In the early 1980s, the "big three" EDA companies (then called computer-aided engineering, or "CAE") were Daisy Systems, Mentor Graphics, and Valid Logic. All derived a considerable share of their revenues from selling workstation hardware. But ECAD was nonetheless consistently profitable, and it went public in 1987.

SDA Systems, like so many other good things in Silicon Valley, started with a dissatisfied engineer. And not just any engineer – Jim Solomon, SDA founder, was a renowned analog engineer with a string of accomplishments at Motorola and subsequently at National Semiconductor. Solomon was frustrated by the lack of analog CAD tools, and he saw the need for a standard format for design data storage.

Solomon wrote a business plan while at National Semiconductor, and while he wasn't originally intending to run a new company, that's what happened. SDA received start-up funding from National Semiconductor, General Electric, Harris Corp., and L.M. Ericsson. The company developed an integrated suite of IC physical design tools. Perhaps SDA's biggest contribution was the idea of a "design framework," developed in co-operation with University of California at Berkeley professors Richard Newton and Alberto Sangiovanni-Vincentelli. The SDA framework provided a common user interface and database, and allowed engineers to integrate tools from a variety of sources.

In 1984, Joe Costello, who would later become the first CEO of Cadence, left National Semiconductor to join SDA Systems as vice president of customer service. In 1987 he became SDA Systems president and chief operating officer. Like ECAD, SDA was a successful software-only EDA company. In September 1987 SDA filed for an IPO, but the October stock market crash derailed those plans.

As the EDA market rapidly expanded in the 1980s, ECAD and SDA realized they could best take advantage of new opportunities by teaming up. In February 1988, ECAD agreed to acquire SDA in a stock swap valued at $72 million. The merger was completed May 31, 1988, and the company was incorporated June 1st as Cadence Design Systems. Huang became a vice president of R&D, Solomon became president of the Cadence Analog Division, and Costello was named Cadence president and CEO.

1989—A Quick Start Out the Gate

1989 was a formative year for the young company in several respects. Cadence completed two strategic acquisitions, launched the Analog Division, and experienced rapid growth, quickly becoming the leading provider of IC CAD tools. Like ECAD and SDA, Cadence continued a "software-only" EDA model, supporting popular third-party workstations and computers.

In March 1989, Cadence acquired Tangent Systems, a provider of timing-driven placement and routing software. The acquisition propelled Cadence to the #1 spot in IC CAD. Tangent's Tangate product became Cadence Gate Ensemble, and Cadence Cell3 Ensemble was an adaptation of Gate Ensemble for standard cell based design. These products took leadership positions in ASIC placement and routing, and they became a major revenue source for Cadence.

In November 1989, Cadence acquired Gateway Design Automation, developer of the Verilog language and Verilog-XL simulation software. The Verilog hardware description language (HDL) represented a new way to do chip and systems design. Instead of drawing gates on a schematic, designers could write code at the register-transfer level (RTL), greatly amplifying productivity.

In the late 1980s, most other EDA vendors were promoting VHDL, an HDL that had Department of Defense (DoD) backing. But Verilog users, already familiar with the C-like language, remained fiercely loyal.

Cadence offered Verilog as an open standard, and today Verilog and its SystemVerilog cousin—both IEEE standards—are far more widely used than VHDL.

Starting with technology initially developed at Harris Semiconductor, the Cadence Analog Division produced Analog Artist, a full-custom IC design software environment that provided schematics and simulation and included a layout editor. A Lisp-based language called SKILL® provided user programmability of the toolset.

Analog Artist set the stage for continuing Cadence strength in the analog IC CAD market. After many years of improvements, Analog Artist evolved into the current Cadence Virtuoso® Analog Design Environment (ADE). SKILL is still widely used to develop process design kits (PDKs), generate parameterized cells (PCells), and interact with and customize tools, including Virtuoso custom/analog IC tools and Cadence Allegro® PCB design tools.

Cadence CEO Joe Costello speaks at the Design Automation Conference in 1997. (Photo courtesy of Design Automation Conference)

Under Costello's charismatic leadership, Cadence grew rapidly during this period. The number of employees went from 433 in 1988 to 978 in 1989. According to Dataquest, Cadence held 44.2% of the $172.3 million IC CAD market in 1989. In 1990, Cadence became the second largest EDA provider, following Mentor Graphics.

Cadence in the 1990s—New Technology and Rapid Expansion

Cadence grew quickly in the 1990s. The growth was fueled by both internal R&D development and a number of strategic acquisitions. In 1991, Cadence acquired Valid Logic, which was then the third-largest EDA vendor in terms of revenues. As a result of this acquisition, Cadence became the EDA revenue leader, a position held for much of the next two decades. While Cadence was already strong in IC design, Valid's area of strength was system design, which involved multi-chip systems and boards. Here are some other key developments that took place in the 1990s:

Continued Development of Analog/Mixed-Signal Offerings

In 1991, Cadence launched the Spectre® simulator, which is still a key offering. This circuit simulator could handle larger circuits than SPICE and run up to 10 times faster. Also in 1991, Cadence brought out the Analog Artist Layout Editor, which linked layout with schematics. Incremental improvements to Cadence analog/mixed-signal products continued throughout the decade, and the Virtuoso name came to identify the Cadence family of custom/analog tools.

Pioneering Work in System-Level Design

System-level design is still thought of as a "new" area in EDA although it's been around for a long time. In the 1990s, Cadence was a pioneer of this technology, then called electronic system design automation (ESDA). In 1993, Cadence acquired Comdisco Systems, which provided a graphical DSP design tool called Signal Processing Workstation (SPW) and a network analysis tool called Block-Oriented Network Simulator (BoNES). Cadence then formed the Alta Group to focus on ESDA.

In 1994, Cadence moved further into ESDA by acquiring Redwood Design Automation, which had developed a system-level simulator. In 1998, the existing Alta products were merged into the Cadence mainstream, and the Felix Initiative was launched to develop an ambitious new level of ESDA tools.

A more immediate step up in abstraction, however, was taking place as engineers made the move from gate-level schematics to RTL design using VHDL or Verilog. Logic synthesis from RTL code to a gate-level netlist was an important part of this methodology. In 1994 Cadence offered a suite of "placement-based" synthesis tools. To bolster its synthesis offerings, Cadence acquired a synthesis startup, Ambit Design Systems, in 1998.

Continuing Innovation in Placement, Routing, and Physical Verification

IC placement and routing and IC physical verification were major Cadence strongholds throughout the 1990s. Cadence continued to innovate in IC physical design in response to rising chip complexity and the move to "deep submicron" designs (meaning process nodes below 1 micron).

Developed at ECAD, the Dracula physical verification product helped Cadence cement its early lead in IC physical design. But in the mid-1990s, Dracula was running out of steam for large designs. In 1995, Cadence introduced Vampire, a hierarchical successor to Dracula that ran 2X to 100X faster.

In 1996, Cadence rolled out Silicon Ensemble for IC placement and routing. In 1999, Cadence brought out Silicon Ensemble Ultra, a next-generation IC physical design solution for 0.18 micron designs.

Bringing Peace to the "Language Wars"

Cadence placed the Verilog language in the public domain in 1990 and a new organization, Open Verilog International (OVI), was chartered to take responsibility for the nascent standard. Cadence continued to provide Verilog simulation tools. Most other EDA vendors, however, were strongly supporting VHDL, and a "language war" between Verilog and VHDL began. Over time, many loyal Verilog users resisted the move to VHDL, and other EDA companies began to provide tool support for Verilog.

In May 1992, Costello gave a keynote speech at the VHDL International (VI) User's Group. He called for an end to the "HDL wars," called on OVI and VI to work together (they later merged to form the Accellera standards organization), and said that Cadence was 100% committed to supporting both languages. In the early 1990s, Cadence offered a VHDL-XL simulator (in addition to Verilog-XL) and then rolled out a new VHDL simulator called Leapfrog.

Cadence entered the formal equivalence checking market with Affirma in 1998. Also that year, Cadence acquired Quickturn Design Systems, gaining the technology that became today's highly successful Palladium® platforms for simulation acceleration and emulation.

Corporate News—A Civil and Criminal IP Rights Case Stuns Silicon Valley

In the early 1990s, an IC placement and routing startup called ArcSys, later re-named Avant!, was challenging established EDA vendors including Cadence. But Cadence executives began to suspect misappropriation of Cadence source code. In December 1995, a police raid on Avant! headquarters in Sunnyvale, California seized potential evidence and kicked off a five-year legal battle among the county, Avant!, and Cadence.

Cadence filed suit against Avant! over the alleged theft of Cadence source code. Avant! countersued, and the two companies went to court many times over the next five years. Eventually, criminal charges were filed against Avant! and several executives as well. At the conclusion of this legal drama in 2001, Avant!'s chairman and six other individuals pleaded no contest to the criminal charges. Avant! was ordered to pay restitution to Cadence.

Finally, in late 2001, Synopsys purchased Avant! for $780 million—at the time, the largest acquisition in EDA history.

In other corporate news, Cadence acquired PCB routing pioneer Cooper & Chyan Technologies (CCT) in 1996. This not only brought Cadence new routing software, but also a new CEO. In 1997, Jack Harding became Cadence's CEO after serving as CEO at CCT. Harding was succeeded as CEO at Cadence in 1999 by Ray Bingham, who had previously been Cadence's CFO.

Cadence in the 2000s—Strengthening Technology, Driving Standards

In the 1990s, Cadence built a solid foundation covering almost every aspect of EDA—including custom/analog design, digital IC design, functional verification, PCB design, and system-level design. In the 2000s Cadence built upon that foundation, and brought forward new technology both from internal R&D and external acquisitions. Cadence also played a key role in EDA standards development, especially with the OpenAccess database, Common Power Format (CPF), and Universal Verification Methodology (UVM).

A Standard Data Model for the EDA Industry

While there have been many EDA standards efforts, OpenAccess may be the most successful and impactful of all. Today, the OpenAccess standard and the reference implementation are widely used by EDA vendors, fabless semiconductor companies, IDMs, and foundries.

Cadence CEO Ray Bingham (center) meets the Queen Elizabeth II in 2000 at the Cadence Livingston Design Center in Livingston, Scotland.

Cadence continues to maintain and upgrade the reference implementation as a service, at no cost to the industry.

The OpenAccess effort began in the 1990s, when large EDA customers—including some who were just starting to buy commercial tools—decided they wanted a common data model and C++ API to provide interoperability among EDA tools. The user companies coalesced into the OpenAccess Coalition under the Silicon Integration Initiative (Si2). When Si2 put forth a request for technology in 2001, Cadence responded by contributing what was then called its Genesis database.

Currently the Si2 OpenAccess Coalition still manages the standard, allowing companies to download the OpenAccess data model, API, and reference database.

Acquisitions Provide Capabilities for Leading-Edge IC Designs

A spate of acquisitions in the 2000s helped Cadence integrate the latest and greatest technology into its IC design tools. They included the following:

- 2001: Purchase of CadMOS brought tools for noise analysis, physical verification, and signal integrity, including the CeltIC® cell-level noise analysis tool. Charlie Huang, present-day senior vice president for Worldwide Field Operations and the System & Verification Group at Cadence, was CadMOS co-founder and CEO.

- 2001: Acquisition of Silicon Perspective included the First Encounter® silicon virtual prototyping tool.

- 2002: Cadence bought Plato, developer of the NanoRoute® system-on-chip (SoC) router that's still in use. NanoRoute technology was integrated into the Cadence SoC Encounter™ product, the forerunner of the present-day Encounter Digital Implementation System.

- 2002: The Simplex Systems acquisition provided advanced technology in 3D parasitic extraction, power grid planning, electromigration, and signal integrity analysis, as well as a highly-respected design services group.

- 2003: Purchase of Get2Chip synthesis startup brought new RTL synthesis technology that formed the basis of today's RTL Compiler. Get2Chip had also developed physical synthesis (integration with placement) technology. Chi-Ping Hsu, Chief of Staff at Cadence, was president and CEO of Get2Chip.

- 2006: Design for manufacturability (DFM) was a big issue at this time, as designers began working at 90 nm and below. Cadence purchased Praesagus in 2006 and Clear Shape in 2007. Both of these DFM companies focused on the impact of manufacturing variability. Invarium, acquired in 2007, provided lithography modeling.

- 2008: Cadence acquired Chip Estimate, provider of chip planning tools and the ChipEstimate.com™ silicon IP portal, which is still heavily trafficked.

Metric-Driven Approach Redefines Functional Verification

In the 2000s, functional verification emerged as a major bottleneck in the IC design cycle. With advanced process nodes, it became possible to place tens of millions of gates on a single chip. Conventional approaches to simulation broke down, and a paradigm shift was needed. Such a shift came about through the Cadence purchase of Verisity in 2005, which brought to a larger marketplace new ideas such as reusable verification methodologies, constrained-random testbench generation, metric-driven verification with functional and code coverage, and use of verification IP (VIP).

Before the acquisition, Verisity was a relatively young EDA company focused exclusively on verification. They had a dedicated verification language, called "e", which is still widely used and was even adopted as an IEEE standard (IEEE 1647). Verisity developed the eRM (e Reuse Methodology), which later provided a foundation for the Open Verification Methodology (OVM) offered by Cadence and Mentor Graphics. OVM, in turn, was the basis for today's Universal Verification Methodology (UVM), which is now supported by all major EDA vendors.

Verisity also provided the Specman® verification environment, which included such features as automatic test generation, data checking, and functional coverage analysis. Verisity pioneered "coverage-driven verification," an approach in which engineers run simulation, collect coverage metrics, and use the metrics to determine whether additional testing is needed. Cadence integrated Specman technology into the Cadence Incisive® verification suite, and refined coverage-driven verification into what is now called "metric-driven verification." In the 2000s, Cadence also moved forward in formal verification. The 2003 purchase of Verplex Systems brought Cadence the widely-used Conformal® product line. In 2005, Cadence released Incisive Formal Verifier, which helps designers verify assertions in RTL code.

A New Way to Describe Power Intent

Low-power IC design emerged as a big concern in the 2000s. Engineers started using low-power design techniques such as clock gating, multiple threshold voltages, power shutoff, and voltage islands. But there was no standard way to specify power intent.

In 2006, Cadence launched the Power Forward Initiative along with Applied Materials, ARM, AMD, Fujitsu, Freescale, NEC, NXP, and TSMC. The organization was chartered to develop the Common Power Format (CPF), which could describe power intent for multiple tools in a single file. In December 2006, Cadence contributed the CPF format to Si2 and in March 2007, the first version of the CPF standard was available to everyone in the industry. It has been successfully used in hundreds of SoC designs.

Cadence competitors led an effort to develop another power format, the Unified Power Format (UPF), which started in Accellera and is now the IEEE 1801 standard. Cadence is actively involved in IEEE 1801 Working Group and is working with customers and other suppliers toward convergence between the two formats.

Continued Improvements in Custom/Analog Design

In the 2000s, Cadence continued to improve its core strength in custom/analog IC design. In 2004, the company acquired NeoLinear, which developed a circuit sizing tool. Tom Beckley, present-day senior vice president for the Custom IC & PCB Group at Cadence, was president and CEO of NeoLinear.

In 2006, Cadence re-tuned the Virtuoso environment to offer a constraint-driven flow and run on the OpenAccess database. This opened the door to the present-day Cadence mixed-signal capability, which uses OpenAccess as a common repository for analog IP design with Virtuoso tools and digital IP design with Encounter Digital Implementation System tools.

Corporate News—Changes at the Top

In 2004, Mike Fister, who had been a senior vice president at Intel, became president and CEO of Cadence. At Intel, Fister was responsible for the Enterprise Platforms Group, and he oversaw the design, development, and marketing of IA-32 processors. Lip-Bu Tan succeeded Fister in 2008, and Tan is the company's current president and CEO. Tan, a respected global venture capitalist, had been a Cadence board member since 2004. While Tan became CEO at a challenging time—in the middle of a recession and anemic EDA market growth—under his leadership Cadence delivered leading-edge technology, forged deep collaborations with customers and ecosystem partners, and experienced 15 quarters of consecutive revenue growth as of October 2013.

Cadence From 2010 to 2012—Advanced Nodes and New Horizons

Taking a view of EDA that goes well beyond silicon, Cadence released the EDA360 vision paper in 2010. The paper was a call to action that emphasized the importance of software applications as a driving force for electronics design. The EDA360 vision includes Silicon Realization, which requires unified flows for analog, digital, and mixed-signal IC designs. This reflects what most people think of as "EDA." But EDA360 also encompasses SoC Realization, which denotes the assembly of complex SoCs using IP blocks, and System Realization, which encompasses embedded software, hardware/software co-development, and PCB and IC package design.

To boost its SoC Realization portfolio, Cadence acquired Denali Software, the major supplier of memory models and IP, in 2010. Today Cadence offers the industry's largest selection of memory models and verification IP (VIP), along with a growing portfolio of high-performance interface IP and memory IP. Martin Lund, formerly senior vice president and general manager of Broadcom's Network Switching Business, joined Cadence in 2012 and is now senior vice president of the IP Group. In 2013, Cadence acquired Cosmic Circuits, a leading provider of analog/

mixed-signal IP; Tensilica, a provider of dataplane processing IP; and the IP business of Evatronix, which includes USB, MIPI, display, and NAND Flash controller IP.

Cadence addressed System Realization in 2011 with the System Development Suite, a set of four connected hardware/software development platforms including virtual prototyping, simulation, acceleration/emulation with the Palladium XP platform, and FPGA-based prototyping.

From a Silicon Realization perspective, Cadence has continued to show leadership in advanced node design for both custom/analog and digital designers. This called for deep and unusually early collaborations with foundries and IP companies. In 2010, the Encounter 9.1 platform added 28 nm support. At the end of 2012, Encounter technology fully supported 20 nm and had been used for two announced 14 nm tape-outs. Likewise, the Virtuoso Advanced Node environment introduced in 2013 supports 20 nm and below with technologies such as automatic color-aware design for double patterning, "partial" layout to get early parasitic estimates, and analysis of layout-dependent effects.

FinFETs represent an exciting new transistor technology that promises tremendous power and performance advantages at 16 nm/14 nm and below. Cadence has been at the forefront of this technology. For example, Cadence helped a team at the University of California at Berkeley develop the BSIM-CMG device model for FinFETs. In 2012 Cadence announced two 14 nm FinFET test chip tapeouts with Cadence tools. In 2013 ARM® and Cadence partnered to implement an ARM Cortex®-A57 processor in a TSMC 16 nm FinFET process.

In 2011, Cadence acquired Azuro, the inventor of "clock concurrent optimization" technology, which represents a paradigm shift in IC physical design that optimizes the clock tree and the logic simultaneously. In 2013 Cadence made a major move into the timing and power signoff market with the Tempus Timing Signoff Solution and the Voltus IC Power Integrity Solution. Anirudh Devgan, formerly an executive with Magma Design Automation, became the senior vice president of the Digital and

Signoff Group at Cadence. Cadence also developed a comprehensive suite of technology to support 3D-ICs, an emerging technology that promises to ultimately allow designers to stack dies using different process nodes. 3D-IC design requires an integrated approach to analog, digital, IC package, and PCB design—and Cadence has all of these technologies.

Summary – The Future Looks Bright (and Very Small)

The Cadence journey has not been without its challenges. But as of this writing, prospects look bright. The company has enjoyed several years of solid growth. Cadence posted $1.326 billion in revenues in 2012, and employed around 5,200 people by the end of that year.

Cadence today is uniquely positioned to partner with semiconductor and system companies. Here are several key strengths:

- End-to-end, integrated (yet open) flows for custom/analog design (Virtuoso products), digital implementation (Encounter products), functional verification (Incisive products), and IC package/PCB design (Allegro products), all of which are time-tested and in widespread industry use.
- Market leadership in verification IP and an increasing portfolio of design IP, with unique offerings in memory, storage, high-speed interfaces, analog and mixed-signal cores, and configurable dataplane processing units (DPUs).
- A deeply connected set of hardware/software development platforms.
- A leadership role in new technologies including 20 nm processes and 16/14 nm FinFETs.
- Strength in analog, digital, and packaging/PCB uniquely positions Cadence for 3D-IC design.
- Deep collaborations with all major foundries, working at the very early stages of process development in many cases.
- Deep collaborations with IP providers including ARM.
- A vision of the future that goes beyond semiconductor design to encompass systems and software.

As we head down the semiconductor process node curve toward 16/14 nm and 10 nm and beyond, it's an exciting time for the electronics industry. EDA made the electronics industry possible, and Cadence contributions will play a major role for many years to come.

Cadence, Allegro, CeltIC, Conformal, First Encounter, Incisive, NanoRoute, Palladium, Silicon Ensemble, SKILL, Specman, Spectre and Virtuoso are registered trademarks and Chipestimate.com and SoC Encounter are trademarks of Cadence Design Systems, Inc. All other trademarks are the property of their respective holders.

In Their Own Words: Synopsys

> *Synopsys has been instrumental in creating and advancing the EDA industry for 25 years. In this section, Synopsys shares their history, technology, and their role in creating the EDA business we have today.*

Synopsys is a market and technology leader in the development and sale of EDA tools and semiconductor IP. One of the largest software companies in the world, Synopsys grew from a small, one-product startup in 1986 to a global leader with more than $1.7 billion in annual revenue in fiscal 2012.

Synopsys' founders, from left to right: Bill Krieger, Aart de Geus, Dave Gregory, Rick Rudell.

In the late 1970s, Dr. Aart de Geus, Synopsys' co-founder, chairman and co-CEO, immigrated to the United States, enrolled at Southern Methodist University in Dallas, and became immersed in the school's electrical engineering program. He soon went from writing programs designed to teach the basics of electrical engineering to hiring students to do the programming. In the process, he discovered the value of taking a technical idea, creatively building on it, and motivating others to do the same.

In 1986, after earning his Ph.D. and gaining CAD experience at General Electric, Dr. de Geus and a team of engineers from GE's Microelectronics Center in Research Triangle Park, North Carolina—Bill Krieger, Dave Gregory and Rick Rudell—co-founded logic synthesis startup Optimal Solutions Inc.

In 1987, the company moved to Mountain View, Calif. and became Synopsys (for SYNthesis and OPtimization SYStems). That same year, Synopsys proceeded to commercialize automated logic synthesis via the company's flagship Design Compiler tool. This foundational technology transitioned chip design from schematic- to language-based. Without it, today's highly complex designs—and the productivity engineers can achieve in creating them—would not be possible.

Early on, Synopsys established relationships with nearly all of the world's leading chipmakers and gained a foothold with its first products. Using synthesis, companies saw they could cut their custom-chip design time by at least 30 percent. By 1992, the same year Synopsys completed its initial public offering (IPO), the company's customer base included nine of the top 10 computer makers, and the top 25 semiconductor companies.

During that time, Synopsys established strategic partnerships with the leading foundries and FPGA companies, acquired some early EDA point tool providers, launched more than two dozen products, and began to build a long-term strategy to integrate EDA and IP. In just six years ,Synopsys had achieved a run rate of $250 million.

The Implementation Revolution

In the 1980s, gate-level entry or schematic capture paired with gate-level simulation, called CAD (computer aided design), was the predominant chip design methodology in use. Although CAD increased productivity, it still required engineers to draw the circuits to be implemented. The introduction of high-level languages (HDLs) like Verilog (1984) and VHDL (1987) justified the creation, development and growth of Design

Compiler. Synopsys' Design Compiler was fundamental in transforming CAD technology into EDA by providing engineers with a vastly more powerful way to develop Integrated Circuits (ICs). Designers could now describe the functions to be implemented in a circuit using an HDL and let Design Compiler derive the required circuitry. The advent of EDA enabled engineers to simultaneously address scale complexity and systemic complexity. By the mid-1990s, Design Compiler had become the de facto standard for RTL logic synthesis, offering a ten times multiple in designer productivity.

As semiconductor manufacturing technology capabilities continued to grow in line with Moore's law, circuit complexity increased. Towards the end of the 1990s, meeting timing in submicron ICs became a major design challenge. Simple wire-load models were no longer able to accurately predict timing. Synopsys took the lead in addressing this challenge, expanding its technology and products from synthesis to all areas of front-end design, including timing, test, and simulation.

In 1997, the company's development efforts in the area of circuit signoff yielded PrimeTime for static timing analysis of gate-level designs. PrimeTime became successful because it offered accurate timing calculations, support for back-annotation, use of standard formats, signoff endorsement from ASIC vendors, superior timing analysis capabilities for debugging circuits, the ability to identify false paths, and more. Through its broad adoption, PrimeTime became the most widely used tool of its type in the industry, and the cornerstone of a complete signoff suite for timing, signal integrity, power and variation-aware analysis.

Also in 1997, Synopsys acquired EPIC Design Technology, a company that had pioneered commercial transistor-level Fast SPICE simulation technology. On the test front, Synopsys was working on a breakthrough that it brought to market in 1999—TetraMAX ATPG (automatic test pattern generation), followed the next year by DFT Compiler, a single-pass test synthesis tool.

Synopsys next set its sights on developing the back-end flow through organic development and acquisition. In executing two of the largest acquisitions in EDA history, Synopsys obtained key additions to its place and route, parasitic extraction, and manufacturing-aware product offerings. In 2001, the acquisition of Avant!, with its advanced implementation tools, helped Synopsys establish its technology more broadly across the overall design flow. More than ten years later, Synopsys acquired Magma Design Automation, whose core EDA products were highly complementary to Synopsys' existing portfolio in IC implementation, as well as in analog custom design.

Beginning in 2000, growing complexity and ever-shrinking process nodes and schedules made it critical to manage design costs while still delivering better results and faster turnaround time. Synopsys began developing a comprehensive, tightly integrated implementation platform. Synopsys' Galaxy Implementation Platform integrated all tools required for physical implementation of an IC into a coherent environment that simplified how engineers move from one tool to another to increase productivity and lower chances of errors. A major component of that platform, IC Compiler, was released in 2005, giving designers a single, convergent, chip-level physical implementation tool that offers benefits such as superior quality of results (QoR), shorter turnaround time, design cost reduction and ease of use.

Don't Trust—Verify

In the mid-1980s, most semiconductor companies and CAD vendors had their own simulators and utilized multiple gate-level languages to describe a circuit and test its functionality. When Gateway Design Automation built the Verilog XL simulator, Verilog became popular because it allowed engineers to describe a circuit at the functional level.

When Synopsys acquired Viewlogic in 1997, it put aside internal development of a Verilog simulator and focused on VCS, a popular, highly competitive Verilog simulator that Viewlogic brought into the fold (Viewlogic had acquired the creator of VCS, Chronologic, in 1994.) Synopsys has since continuously improved VCS, boosting its performance by at least 2X with each new release.

In the early years of the new century, the use of Verilog increased due to its simplicity. As design complexity increased, Verilog was showing its limitations but the market did not show any sign of returning to a greater use of VHDL. Too many college graduates had been trained in the use of Verilog and changing to VHDL, albeit a more powerful language, would have been too expensive. Synopsys proposed and provided leadership for a project, later adopted by the IEEE, to expand the capabilities of Verilog. Thus SystemVerilog was born and standardized.

Increased complexity combined with the need for hardware/software co-development required the industry to expand the definition of high-level design, including adopting the use of multiple modeling languages, such as C++, to complement SystemVerilog. As a result, Synopsys evolved its vision of simulation to emphasize a verification platform and began to develop the adjacent technologies of coverage, testbench and formal verification using assertions.

This vision became "smart verification" in 2002, with Synopsys building the Discovery Verification Platform, a unified environment with all the adjunct technologies internally developed. Building this unified environment allowed Synopsys to combine system-level verification, HDL simulation, mixed-signal simulation, testbench automation and functional coverage on a single platform. Synopsys had begun working to apply formal techniques to complement verification problems in 1997. Hybrid formal verification was a new approach to functional RTL verification, combining a formal property-checking capability with the VCS Verilog simulator. Synopsys' hybrid formal verification solution, Magellan, was launched in 2003 as part of the Discovery Verification Platform.

Low power became a more dominant factor in mid-2005 with the emergence of mobile technologies and their attendant power conservation requirements. To address power management design challenges, Synopsys acquired ArchPro Design Automation, whose technologies enabled engineers to address power management challenges in multi-voltage designs from chip architecture to RTL and gate-level design. With this acquisition, Synopsys integrated low power verification techniques natively into VCS.

With increasing verification efforts, users needed not only tools for design verification, but also building blocks, or IP. Verification IP (VIP) tools and good verification methodologies are now essential. To address this need, Synopsys increased its investment in these areas. The Discovery VIP suite, introduced in 2012, is SystemVerilog-based and features native support for industry standard verification methodologies, including UVM (Unified Verification Methodology). The methodology includes key building blocks used in verification, while the VIP includes basic protocols/models to validate the behavior of implemented blocks of IP.

In the early 1990s, only a few EDA companies specialized in hardware-based circuit emulation, and Synopsys was not one of them. The need for emulation was growing as design size and complexity increased and verification needed to be run at higher speeds. An early effort to enter the emulation business with the purchase of Arkos Design Systems in 1995 ended when Synopsys divested itself of the company in 1997. Synopsys continued to investigate avenues for reentry, and in 2012 acquired emulation leader EVE. During the same year, Synopsys also acquired SpringSoft, which had the widely used simulation-independent verification debug tool, Verdi. With these additions to its portfolio, Synopsys could offer a complete verification environment, combining dynamic and static verification, emulation and debug, and advanced prototyping capabilities.

Leading the Way in IP

By the early 1990s, Synopsys understood that IP blocks were an integral part of EDA. Establishing itself early in the market, the company began building an IP portfolio through both organic development and acquisition. Since the beginning the focus has been clear: enable designers to meet their time-to-market requirements and reduce integration risk by providing the high-quality IP they need, when they need it.

The DesignWare family, first launched in 1992, offered a collection of technology-independent, reusable building blocks such as adders and multipliers. DesignWare freed engineers from designing the same logic circuits for every design. As synthesis technology advanced through the years, complex IP blocks were added to the library, e.g., 8-bit microcontrollers, AMBA on-chip bus IP, and Verification IP (also known as SmartModels, from Logic Modeling). With these additions, the product became known as the DesignWare Library—and it has been the most widely used library of foundation IP ever since.

Fast forward a decade to the new millennium. An explosion in the usage of standards-based communication protocols set the stage for the emergence of the commercial IP industry as companies realized they needed to focus their efforts on the differentiated portions of their design and not on developing standards-based IP. In 2002, Synopsys acquired inSilicon, adding popular interface protocols such as PCI-X, USB, IEEE 1394, and JPEG to its DesignWare IP portfolio. By acquiring Cascade Semiconductor in 2004, Synopsys rounded out its already successful DesignWare PCI Express Endpoint solution with root port, dual mode and switch ports, providing designers with a complete high-performance, low-latency PCIe IP solution. Also in 2004, the acquisition of Accelerant Networks brought serializer-deserializer (SerDes) technology to Synopsys.

In 2009, Synopsys moved into the leadership position in the analog IP business with the acquisition of the Analog Business Group of MIPS Technologies. The acquisition added a new family of analog IP to the DesignWare IP portfolio, including analog-to-digital converters (ADCs), digital-to-analog converters (DACs), and audio codecs.

The acquisition of Virage Logic in 2009 brought logic libraries and embedded memories into the fold, enabling designers to achieve the best combination of power, performance, and yield; memory test and repair; non-volatile memory; and ARC processors targeted at embedded and deeply embedded applications. Throughout 2010, Synopsys continued to introduce new products that would help designers integrate advanced functionality into their SoCs.

In 2011, the focus became helping designers develop 28-nm SoCs. With this process, and with each process node to follow, IP became more foundry-dependent. Synopsys announced the availability of DesignWare Interface PHY and Embedded Memory IP for TSMC's advanced 28-nm process, as well as its collaboration with UMC on embedded memory and logic library in 28-nm. The next year, designers started to integrate more and larger third-party IP into SoCs. It wasn't enough to just provide individual IP blocks, the market needed complete IP subsystems to ease the integration effort. Accordingly, Synopsys released the industry's first 28-nm Multi-Gear MIPI M-PHY IP supporting six standards. The shift to IP subsystems, 20-nm IP and FinFET increases the need for an IP provider that can support key technology advancements and strong foundry relationships.

Synopsys became the industry's trusted IP partner by prioritizing the top five customer criteria for selecting an IP provider: IP technology leadership; quality/silicon-proven IP; market leadership; brand reputation; and breadth of IP portfolio.

Prototyping—Knowing You Are Building the Right Thing

As Synopsys built its IP portfolio, it recognized the growing importance of high-level synthesis and embedded system-level design. The company was quick to identify the trend toward advanced prototyping technology, including virtual prototyping and FPGA-based prototyping for hardware/software co-design.

For several decades, prototyping of systems has been a crucial part of product development cycles. There are two major prototyping methods: one is virtual prototyping or, as many people call it, system simulation. The other is hardware prototyping, which involves physically building a close enough approximation of the real system. In both cases the goal is to observe the behavior in a way that allows engineers to confirm that the actual system being built will work as intended once the product is manufactured.

Synopsys started its involvement in the prototyping market with the acquisition of COSSAP in 1994. At that time many companies were developing chips for communications systems-either mobile cellular, satellite or wired. It was very expensive to build an entire prototype in hardware to verify if the system provided enough performance to transmit voice and data properly. Until the introduction of hardware emulators, prototyping a new system had two major drawbacks. First, building a prototype was both costly and time-consuming. It meant designing and building a hardware system that worked as reliably as the intended product. Second, as semiconductor technology developed, the operational throughput of a system built with discreet parts was different than what it would be in silicon, rendering some of the outcomes of the prototyping misleading or irrelevant.

Concurrently through the early 2000s, the complexity of RISC processors—the CPU architecture of choice for communication applications—grew along with the chip content, which now included more interfaces that needed new device drivers. As a result, electronics companies faced the problem of developing increasingly complex software. Since the late 1960s developing firmware for a processor meant building a computer model of the processor so that software could be executed and results observed. This technology became known as virtual prototyping. As the decade progressed, three different startup companies pioneered the commercialization of generic virtual prototyping systems. Synopsys eventually acquired all three companies: Virtio (in 2006), VaST and CoWare (both in 2010). The technologies of all three were fully integrated into one product, Virtualizer, that allowed users to develop and debug software before the actual hardware was available. This capability is the foundation of hardware/software co-development.

The most common method in building a hardware prototype is to use FPGAs to model hardware destined to be implemented in silicon. Synopsys had identified a market need for scale and cost reduction through the reuse of off-the-shelf infrastructure built on top of a very robust tool flow for prototyping. To address this need Synopsys acquired Synplicity in 2008, with its High-performance ASIC Prototyping System (HAPS) solution, and the CHIPIt technology from ProDesign, which together provided scalable technologies for this purpose.

Another major requirement to improve the efficiency of virtual prototyping was the availability of a standard modeling language. Back in 1999, Synopsys organized an industry consortium (Open SystemC Initiative) with major EDA vendors and electronics companies to define a common language for IP modeling. Synopsys offered its technology, as did CoWare and others, to the consortium. The result was SystemC, a language based on the popular C programming language. Since then Synopsys has continued to play a leadership role in extending the capabilities of virtual prototyping tools, including the SystemC TLM (transaction level modeling) 2.0 interface. The IEEE has standardized both the SystemC language and the TLM 2.0.

Partnering for Success

In all of its development activities, Synopsys works closely with customers to formulate strategies and implement solutions to address the latest semiconductor advances. This close collaboration with customers is one reason Synopsys was effective in creating a successful services offering.

Synopsys was born during the ASIC "revolution" of the 1980s as design teams at companies like Sun and Motorola raced to avail themselves of the swelling cell capacities of cell-based and gate-array designs. At that time, handoff to each ASIC supplier was complete only after running gate-level simulation using the ASIC vendor's own proprietary timing calculator.

Through customer insistence, Synopsys was able to cement broader collaborative relationships with the two largest ASIC suppliers at the time, LSI Logic and VLSI Technology, despite both companies' reliance on their own internally-developed EDA tools. Eventually, most ASIC vendors would further expand their support and use of Synopsys Design Compiler synthesis and optimization in their design centers. In the summer of 1989, eight ASIC vendors supported Synopsys synthesis. By the summer of 1991, 27 ASIC vendors supported Synopsys synthesis, with 20 using Design Compiler in their own design centers.

Adding an ASIC business meant semiconductor companies had to "externalize" tool flows and cell libraries so that they worked with external tools like Design Compiler, thus opening the door for internal IDM use as well. Developing and supporting internal tools was a very costly task. When the IDMs saw the growing productivity achieved using commercially available synthesis and optimization tools, they began to adopt them for internal development. By the mid-1990s, many of the IDMs' internal design teams had broadly adopted Synopsys' synthesis and optimization design flows.

In the early 1990s, Synopsys' ASIC flow grew to include test synthesis and VHDL signoff. Synopsys solicited each ASIC vendor to support the full flow, but the new capabilities were a challenging sell. While the ASIC vendors loved the concept of manufacturing test and ATPG automation, test vectors other than the end customers' "signoff vectors" complicated the ASIC vendor's business model and added practical problems such as a lack of scan-ready testers. Thanks to a combination of factors, including end-customers who really wanted ATPG and VHDL signoff, and substantial market education, the full flow gained traction.

By late 1995, 38 ASIC vendors and 11 FPGA vendors supported Synopsys synthesis, with most supporting test synthesis and VHDL signoff as well. In the mid-1990s new players entered the ASIC market. TSMC entered the market, offering cutting-edge 0.5 μm standard cells and gate arrays. Two large IDMs, Samsung and IBM Microelectronics, also began to highlight their offerings, with IBM promoting a market-leading 0.35 μm, 1.6 million gate capacity gate-array. All three worked closely with Synopsys to meet their unique flow, timing calculation and test requirements. All three would go on to become long-term pillars of the newly emerging fabless/foundry market.

As the 1990s advanced, the ASIC semiconductor world began to turn on its head. There had always been a number of systems and semiconductor companies, like Chips & Technologies, Xilinx, and Altera, that subscribed to a fabless model where they supplied their finished GDSII layouts to another semiconductor company who would fabricate the devices for them. But in the mid-1990s, driven both by cost consideration and the introduction of pure-play foundries like TSMC, and by the growth of fabless semiconductor startups like Broadcom and Qualcomm, former ASIC users and new startups began to transition to the fabless model.

A new set of horizontal, multi-foundry IP suppliers, including Artisan Components and Virage Logic, emerged to offer standard cells and memories to companies that had traditionally relied upon the ASIC vendor to provide them. Synopsys quickly developed relationships with these budding new IP suppliers to deliver ASIC-like flows for fabless end-customers. In 2000, Synopsys and TSMC collaborated to develop the very first ASIC-like foundry reference flow, TSMC Reference Flow 1.0, which proved a huge success for end-users transitioning to a foundry flow.

One other important transition had begun in the IP market. Formerly, nearly all the significant IP offered by ASIC vendors and the nascent third-party IP market, except for Synopsys DesignWare, was delivered as hard IP. Synopsys, through a couple of prior attempts to productize significant digital IP blocks, unlocked the RTL coding and implementation methodology needed to deliver consistent results from soft, RTL-based IP. Fortuitously, this happened at about the same time ARM was looking for a better way to implement its popular ARM7TDMI processor with multiple ASIC vendors and foundries. Synopsys collaborated with ARM to develop synthesizable RTL versions of its ARM7 and ARM9 processors and the first reference flow—an accompanying Galaxy Implementation Reference Methodology ("iRM") that delivered the flexibility of soft IP with the performance, area and predictability of hard IP. ARM proceeded to create its subsequent processors in synthesizable RTL form and collaborated with Synopsys to deliver iRMs for them. The RTL IP design and delivery methodology was captured in the popular Reuse Methodology Manual [Springer], which is still in broad use today.

Synopsys' collaboration with key partners including ARM, TSMC, Samsung, GLOBALFOUNDRIES, and UMC, and with leading mixed-signal foundries like TowerJazz, Dongbu, and MagnaChip has illuminated the development path to tools and methodologies for the next process node and toward the next level of designer productivity.

Complementary Acquisitions

The combination of in-house technology innovation and strategic acquisitions helped drive Synopsys' success as the company extended beyond its core business to address emerging areas of great importance to its customers. The complementary acquisitions of Avant! and Magma are the two most significant examples. Other significant acquisitions have included Viewlogic, Synplicity, Virage Logic, EVE, and Springsoft.

As challenges associated with analog/mixed-signal (AMS) design escalated, Synopsys integrated several companies with complementary technology to address various AMS design aspects. These included Nassda

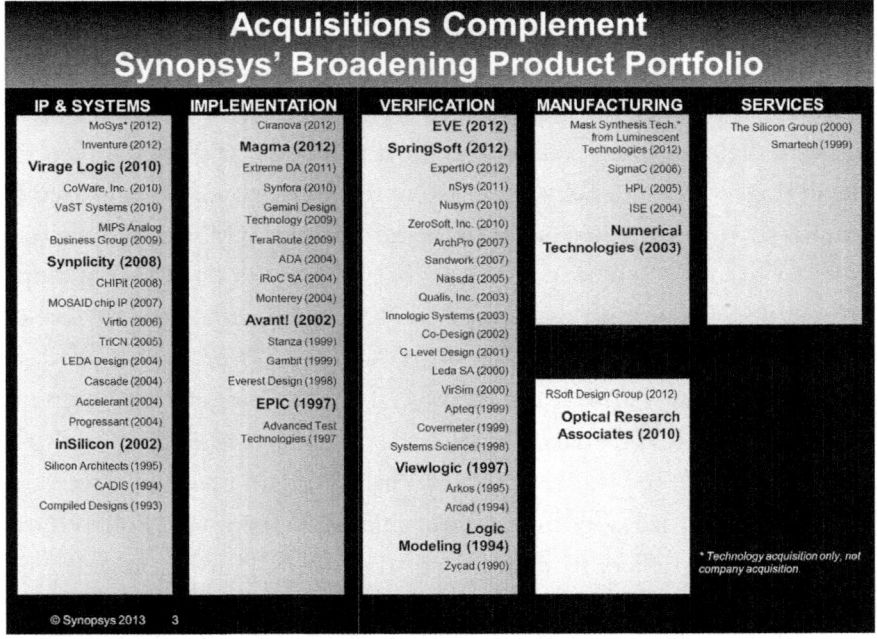

(AMS simulation), Sandwork (AMS verification), MIPS Technologies' analog IP group, and two companies with offerings in custom analog design: Ciranova and SpringSoft. SpringSoft also had a strong offering in verification, adding to a history of successful Synopsys acquisitions in this space.

Synopsys' purchases of Avant!, SIGMA-C and ISE brought TCAD tools

into the fold. Mask synthesis and data prep, via the purchase of Numerical Technologies and Luminescent Technologies, also became important additions to Synopsys' manufacturing tool offering, as did solutions for the design and analysis of high-performance, cost-effective optical systems. To this end, Synopsys acquired Optical Research Associates, a leading provider of optical design, analysis and modeling software, and RSoft Design Group, a maker of photonics design and simulation software.

Diverse Leadership

The most successful companies have a strong team and solid leadership at their core. Over the years, Synopsys assembled a team with diverse global backgrounds and many decades of combined semiconductor industry know-how. The company's co-CEOs embody this diversity and expertise.

Synopsys co-CEOs Aart de Geus and Chi-Foon Chan.

Dr. Chi-Foon Chan had been Synopsys' president and chief operating officer since 1998, and joined Dr. de Geus as co-CEO in 2012. Dr. de Geus and Dr. Chan maintain an effective partnership that recognizes the breadth and complexity of Synopsys' business.

Dr. de Geus has a philosophy: If something already has value, how can it be moved to the next level? It was this approach that essentially informed the discovery of how fostering talent, technology and education can yield exciting results that drive ongoing innovation. One can only imagine what future years will hold.

Chapter 7: Intellectual Property

The current state of innovation in electronics wouldn't be possible without one very important enabling technology: semiconductor intellectual property, or just IP. Where once, all the components of an electronic product—the microprocessor, memory, audio/video encoders, IOs, and other functions—existed on separate chips, today all those functions are integrated onto single SoCs.

SoCs have become more valuable than single-function chips because they offer better performance and more functionality in less space and with less power. They also created efficiencies in manufacturing and packaging that make them attractive to the bottom line. They were made possible by the development of the IP business model.

Although IP often refers to the general intellectual property of a business, such as trademarks, patents and copyrights, in the context of the semiconductor business it usually refers to semiconductor intellectual property, or SIP. However, this is usually just called IP, as it is in this book.

How the IP Business Developed

The development of the IP business was predicated on the changes related to the growth of the ASIC design business, the fabless model, the EDA industry, and pure-play foundries. Recall that many companies that made electronic systems did the front-end design themselves, and then passed the design off to an ASIC company (such as VLSI Technology,

LSI Logic, or IBM) for physical implementation and manufacturing. But in the 1990s, systems companies began to change their design methodology and business models to one in which they did all the design, from concept to tape-out, and hired a foundry for the manufacturing (and usually the packaging and testing of the chip too). This made systems companies and semiconductor companies start to look pretty similar.

Three factors drove this change: the first was the ready availability of effective physical design tools; the second was growth of wafer manufacturing services from pure-play foundry companies such as TSMC. However, for this new model to work, a third key element was needed—access to standard cell libraries and memories, the basic building blocks of any design. Historically, semiconductor companies designed these elements themselves; every company with its own standard-cell libraries, every company with its own memories. However, no semiconductor company differentiated itself by the quality of its standard cell libraries or memories. At the same time, these IP require huge amounts of work to design and maintain, especially considering that they must all be revised constantly as the manufacturing process changes. As soon as the economic downturn in the early 1990s hit, semiconductor companies decided that keeping a huge internal group of people just to develop libraries and memory didn't make sense.

Systems companies didn't have the knowledge in-house to make their own IP, and the semiconductor companies that once made and licensed IP were quickly ditching their IP groups. There was suddenly a clear need for a new type of company, one that specialized in creating standard cell libraries, and an available pool of talent to do it. At this point, as system companies began to take charge of the entire design flow, their demands for libraries and other IP drove the rapid growth of the IP business. The niche was filled by companies such as Compass Design Automation, Artisan, and Virage Logic. Although this was technically semiconductor IP, this name was not yet in use. It was simply known as the library business, to distinguish it from other parts of EDA.

In the early 1990s, though, the libraries available were specific to a given foundry. At first, the foundry's only business was manufacturing chips. It soon became clear that the timely availability of high quality standard cell libraries and other IP was an important enabler for the foundry business. This realization led to a key change in the IP business model in the late 1990s when Artisan made a deal with TSMC that gave designers free access to Artisan libraries if they used TSMC as their foundry. Artisan changed from an upfront licensing model to a royalty model backed by the foundries and bundled invisibly into the wafer price. The standard cell library that once cost $1 million was now free to customers, with a royalty paid to the IP company by the foundries based on wafer sales. This new business model gave IP companies like Artisan and Virage very healthy valuations.

But as SoCs got larger over time, systems companies needed more than just cell libraries; they needed other functional blocks, like processors. The processor IP business was largely born through the collaboration between Apple and Acorn Computer (often called the British Apple) in the late 1980s. That collaboration gave birth to ARM, which has since grown into the most successful IP business, and their processors are in almost all modern mobile electronics.

When it was clear that the Newton was not going to be the big success that had been anticipated, ARM began to license their microprocessor to all-comers. The timing was right, as more system companies began making their own chips, and more second-tier semiconductor companies also needed microprocessors. The success of ARM was ensured when cell-phone companies standardized on the ARM7TDMI products as the control processor for the second generation of mobile phones.

Once standard cell libraries, memories, and microprocessors were widely available to be licensed by anyone, it suddenly became possible, both from a business and a technology view, to design more complex chips that could then be manufactured at a pure-play foundry or within an IDM's own foundry.

SOCs Increase the Need for IP

As SoCs added more functionality, there was a need for more IP, such as USB or PCI interfaces. Because these interfaces are defined by industry standards, there is limited opportunity for companies to differentiate by designing a "better" interface block. About this time, in the late 1990s, these types of blocks were starting to be called IP. The barriers to entry into this IP business were low, partly because it required only a few designers who know how to design an interface block. In the late 1990s, literally hundreds of small IP companies were born, many only supplying a handful of interface elements.

This level of competition drove prices down. Originally, IP was sold on the basis of an up-front licensing fee and a back-end royalty on the manufactured parts paid by the foundry. As licensing prices came down, and royalties were reserved for only the most exclusive IP blocks such as microprocessors, most of the smaller companies failed. It became obvious that success in the IP market meant having a broad portfolio. A company with just a handful of blocks was doomed.

All kinds of companies entered the IP business, including EDA companies. Mentor Graphics put together a large portfolio of IP through a mixture of acquisition and internal development, but was never highly successful in the IP business and eventually exited the IP market only to re-enter it a few years later through acquisitions. Synopsys had some IP called DesignWare that was initially focused on adding higher-level blocks such as adders and multipliers to their synthesis tools and methodologies. They gradually expanded their portfolio through a mixture of acquisitions like Virage Logic and internal development, and today are the #2 IP supplier behind ARM. Cadence followed suit with the acquisitions of Denali, Tensilica, Cosmic Circuits, and others. A complete list of IP acquisitions by EDA companies is available on SemiWiki.com.

Ironically, a key catalyst for the IP business, industry standards, also made it very difficult to turn a profit in the non-differentiated IP market. PCIe, DDRx, USB, MIPI, and other interface IP tends to be non-differentiated because the specification is fixed by industry standards. They should all be identical apart from cost or quality. These IP made the development of SoCs faster and cheaper. But, for standards-based IP, there was no way for suppliers to demonstrate how they were better than their competitors. Potential customers were just as likely to make the low-value part themselves as to buy it, which kept prices too low. There was also very little reason for repeat customers, unlike in EDA. For example, if you used the place and route tool from Synopsys for 90 nm, there was a good chance you would use the Synopsys tool for 65 nm because of the costs associated with changing design flows and retraining engineers. But if you purchased a USB1 interface from one company, there was no good reason to believe that they would offer the best choice for USB2.

The companies that were most successful in the IP industry were those that sold something that was not easy to do. There are three especially notable sub-segments of IP where companies managed to find success: microprocessors, on-chip communication architectures, and analog IP.

The first class of high-value IP, the microprocessor, is more than just a structure that you put on silicon. It requires compilers, debuggers, in-circuit-emulators, operating systems, and more—a complete ecosystem to surround it. This surrounding ecosystem is the barrier to entry for the microprocessor IP business, not the difficulty of designing a microprocessor in silicon. ARM has been the most successful at this. ARM transformed the microprocessor market by adopting a balanced business model of up-front license fees, royalty revenue from every chip sold by customers incorporating ARM IP, and revenues from related development tools and customer support. ARM also acquired Artisan Libraries in 2004 and the resulting ARM ecosystem is second to none.

ARM wasn't the only important microprocessor licensing company, though. Two especially notable ones were US-based MIPS, a spinout from Silicon Graphics that had a processor that featured higher performance, but also higher power usage, than the ARM architecture, and UK-based Imagination Technologies that licenses a whole range of processors, but most notably graphic processor units (GPUs). The MIPS processors found success in both the television set-top box market (and later in digital video recorders, DVRs), in the video-game console market, and also in desktop printers.

However, as the cell-phone market grew, MIPS decided that the low margins on phones didn't justify entry into the mobile processor business. As the transition from feature phones (dumb phones) to smart phones took place, Imagination's PowerVR GPUs were in many smart phone designs, most notably the iPhone. In 2013, Imagination completed the acquisition of the operating business and selected patent properties of MIPS for $100 million. This immediately stabilized MIPS, which was having difficulty attracting any new accounts because its future was so uncertain. Because Imagination Technologies is a global leader in multimedia and communication technologies, this seems to be an excellent fit for the MIPS CPU architectures and IP cores.

There are other companies licensing microprocessors too. Synopsys, through the acquisition of Virage in 2010, has its own microprocessor architecture, called ARC. In digital signal processing and other specialized dataplane applications such as audio, there are CEVA and Tensilica (acquired by Cadence in 2012 for $380 million). All of these companies have put together the ecosystem of tools and software that is needed over and above just the semiconductor IP itself. These microprocessors have all shipped in the billions of units, some of them in tens of billions of units.

The second class of high-value IP is communication architecture for use between blocks on a chip. The two main companies in this space are Sonics, Inc. and Arteris, Inc., both of whom have network-on-chip (NoC) architectures. Again, the investment to produce a general purpose NoC is too much for any system company to undertake on its own; the technical expertise and the amount of software and verification data necessary is a very high barrier to entry.

Finally, there is analog IP, which doesn't suffer from the make-versus-buy problem because most system companies don't have the capability to design, say, a DDR PHY (the physical interface that ties an SoC to its memory subsystem). The IP company Denali specialized in just these sorts of memory interfaces. Denali was acquired by Cadence in 2010 for $315 million. As process technology marches on, analog design gets more difficult, and ever more valuable. Those companies with the capability to execute should continue to thrive.

There are of course, other factors in the development of the IP business, including the ability to quickly migrate IP from one manufacturing process to another, and the advent of language-based synthesis that further enabled the quick transition of IP to new process nodes and cell libraries. Unfortunately, we don't have room here to cover all the companies, technologies, clever business strategies, and lucky coincidences here.

In the following chapters, two of the top IP companies, ARM and Imagination Technologies, describe their histories and their roles in the development of the IP business.

In Their Own Words: ARM

ARM, Ltd. is synonymous with IP. They've done more to shape the industry into its current state, and to enable the growth of modern electronic gadgets than any other IP company. In this section, ARM tells their story.

It was on the 26th of April, 1985 (at 3pm to be precise) that the very first ARM silicon sprang in to life—it was a 25K transistor design implemented in 3 μm technology with just 2 layers of metal.

However back then the "A" in ARM stood for Acorn—ARM, the company, had yet to be formed. Acorn sold computers to schools and so cost was a prime concern, this meant that when it came to replace the aging 8-bit 6502 in the BBC Micro with a more powerful microprocessor it had to be cheap.

The BBC Micro

Unfortunately the commercially available alternatives at the time were simply not cheap enough and so Hermann Hauser, the Managing Director of Acorn, decided that Acorn should build its own 32 bit microprocessor. However, he gave the ARM design team two distinct advantages over other microprocessor design teams—no money and no people! So the design had to be simple and straight forward, indeed the first ARM reference model was written in just 808 lines of Basic.

Interestingly, although the ARM silicon worked first time, it appeared to be consuming no power at all, at least, that is what the ammeter said. It turned out that the test board had a fault, which meant the ARM core was effectively unpowered and was running solely on leakage from the I/Os. This low power consumption was a valuable side effect of making the ARM core cheap and turned out to be the key to its success in the emerging mobile electronics market.

1990: ARM Ltd. Founded

In early 1990, Apple was developing a "Personal Digital Assistant" called Newton and was looking for a low power processor to power it. Apple was very interested in ARM but was reluctant to base a product on Acorn's IP and so ARM Ltd. was founded on the 27th of November 1990 as a joint venture between Apple, Acorn, and VLSI Technology.

The ARM Barn

The first office was in a beautiful 17C converted barn just outside Cambridge UK. Apple put in £1.5M of cash, Acorn put in the 12 engineers who had worked on ARM, and VLSI provided the design tools. They also became ARM's first licensee, although that word wasn't used since there was originally no plan to license the core widely.

ARM then set about extending the architecture to meet Apples requirements for 32-bit addressing and endianness support. In January 1992, the ARM610 was complete and Newton was launched in 1993.

Partnership Model

Unfortunately, the Newton was not a great success and so Robin Saxby, ARM's CEO, decided to grow the business by pursuing what we now call the IP business model, which was rather unusual at the time. This was still at the stage that a microprocessor was a whole chip, or most of it, and before it was small enough to just form part of an SoC, although that would change quickly due to Moore's Law.

Later in 1992, GEC Plessey Semiconductors (in the UK) and Sharp (in Japan) became the first two licensees. The following year they were joined by Cirrus Logic and Texas Instruments, the first US licensees.

The ARM processor was licensed to many semiconductor companies for an upfront license fee and then royalties on production silicon. This effectively incentivized ARM to help its partner get to high volume shipments as quickly as possible.

One interesting feature of this IP licensing business model is that the pipeline is very long—it can take years from the time a license is signed until the royalties really start to kick in.

The Apple Newton

When ARM started, it had an internal software group to produce the compilers, assemblers and debuggers that it required. ARM could not do everything necessary internally however. It was still a small company with a niche processor architecture, and companies like Wind River that produced real-time operating systems needed to be paid to port their product lines and support the architecture.

As the ARM architecture became more and more widely licensed, ARM put a lot of effort into building a partner program so that anything that an ARM licensee might need would be available from an ecosystem of 3rd party suppliers. The economics for the partners changed so that gradually selling into ARM's base of licensees was a huge opportunity and nobody needed to be bribed to support the architecture.

1994: "Thumb"—the Big Break

In 1993, Nokia approached TI to produce a chipset for an upcoming GSM mobile phone and TI proposed an ARM7 based system to meet Nokia's performance and power requirements. Unfortunately, Nokia rejected the proposal as the memory footprint of a an ARM7 based solution made the system cost too high—being a 32-bit processor, every instruction took 4 bytes. ARM came up with a radical idea to create a subset of the ARM instruction set that required just 16 bits per instruction. This improved the code density by about 35% and brought the memory footprint down to a size comparable with 16 bit microcontrollers.

Thumb, as it became known, was a major breakthrough that won Nokia over, and arguably is what got ARM in to mobile phones. The first ARM-powered GSM phone was the hugely popular Nokia6110. The ARM7TDMI that powered it went on to become one of ARM's most successful products with more than 170 licensees who have shipped over 10 billion units since its introduction in 1994.

ARM's timing turned out to be very fortunate. The ARM7TDMI was released just as the cell-phone market started its explosive growth. ARM became the standard processor in mobile, as it still is today. Not only did this mean that a lot of cores were shipped, it meant that every semiconductor company needed an ARM license if they were going to participate in selling semiconductors into the cell-phone market.

One licensee of the ARM architecture was Digital Equipment Corporation (DEC). They didn't license any particular core, they had an architectural license and they built their own version of the core on their own process, highly optimized for even lower power and higher performance. It was called StrongARM and debuted in 1995. But there is an interesting twist: many of the team members who developed it moved on when that part of DEC was acquired by Intel (who renamed StrongARM to Xscale, and eventually sold their whole communication business to Marvell). They would go on to create PA Semiconductor, which designed very low power PowerPC cores. Apple acquired them in

2008, took out an architectural license from ARM, and today they form the core of Apple's processor design team working on...ARM cores. In 2013 they produced the first 64-bit ARM inside the Apple A7 that powers the iPhone 5 and the iPad Air.

By the end of 1997, ARM had grown to become a £27 million business with a net income of £3 million and so it was time to float the company. On April 17th, 1998, ARM completed a joint listing on the London Stock Exchange and NASDAQ with an IPO at £5.75. The stock soared and ARM became a billion dollar company almost overnight.

The Move to Synthesizable Cores

Chips were now large enough that a microprocessor only occupied a small part of the chip and so it was possible to build software-based systems on a single chip, the so-called SoC. Microprocessors were one of the first parts of SoCs to use the IP business model since most design teams didn't have the knowledge or desire to build their own microprocessor and certainly lacked the skills to build the tool-chain of compilers and debuggers necessary to make it usable. As a result, ARM was designed into more and more SoCs, especially in the explosively growing cell-phone market where ARM had gradually become the de facto standard.

The Nokia 6110

However, the ARM core was technology-specific "hard IP" and it soon became clear that porting it to so many different technologies was a real bottleneck and that something had to change. A synthesizable core was required that could be licensed to anyone without needing a technology-specific port of the core.

In 2001, the ARM926EJ-S was announced. It was fully synthesizable with a 5 stage pipeline and a proper MMU, as well as hardware support for Java acceleration and some DSP extension. It went on to be licensed by over 100 silicon vendors worldwide and has shipped over 5 billion units to date.

2001 was also the year that Robin Saxby, the original CEO of ARM when it was spun out of Acorn, passed the torch to Warren East who became the new CEO.

2004: Artisan

Artisan Components was a company that designed and marketed standard cell libraries, memory compilers and interface components. These are the basic components of any synthesizable design, the Lego bricks out of which complex designs are built. In 2004 ARM acquired Artisan and so added a physical IP business line.

In recent years, the physical IP business has also been extended to add special cells called Performance Optimization Packs (POPs) that further optimize the process of synthesizing ARM cores for particular processes, most notably for the big foundries that actually manufacture many of the ARM-based designs.

2005: Cortex

The subsequent development of ARM9 and ARM11 families had extended the capability of the ARM architecture in the direction of higher performance with the introduction of multi-processing, SIMD multimedia instructions, DSP capability, Java acceleration, etc. However there were other, potentially much larger market segments, which these processors did not address. So, in 2005, ARM introduced a change of direction and the ARM architecture was split into three "profiles," the upwards and to the right path continued with the Cortex-A, a new range of high-performance real-time processors was introduced as Cortex-R whilst the Cortex-M profile targeted microcontrollers.

2008: Multicore

By 2008, the smartphone market was booming and the demand for increased performance, while at the same time maintaining a long battery life, presented quite a challenge. ARM responded with the Cortex-A9 MPCore, a multi-core processor which was better able to address the huge dynamic range in processing power from idle or playing music to full bore 3D gaming. This was further improved with the introduction of the heterogeneous "big.LITTLE" approach in 2011, which provides high performance with a powerful core when required and then switches back to a lower performance but much lower power core when high performance is not needed.

In a smartphone or a tablet there are two main processors: the application processor, which was already dominated by ARM, and the graphics processor, a specialized core that drives high-resolution screens and is required to run videos and games on such devices. In 2008, ARM introduced their Mali graphics processing unit (GPU). Like the ARM processor cores before, Mali would go on to become the world's most widely licensed GPU architecture.

In 2011, ARM announced the ARMv8 architecture, which took the architecture up to 64-bit without losing backward compatibility with all the existing 32-bit software. This was targeted at expanding ARM's market into the datacenter. A large part of the cost of a datacenter is electricity to power all the computers and cooling to get all the heat out again. ARM's low power design is very attractive and offers the possibility of delivering lots of processing throughput, at low power, low cost and low physical volume.

Today

In July 2013, Warren East retired as CEO of ARM and Simon Segars took the post. Of course, the product line continues to develop with a spread from the Cortex-M0 microcontroller all the way up to 64-bit multicore processors aimed at the corporate datacenter, and with cores in between targeted at attractive large markets such as low-end low-price smartphones.

ARM is now the standard microprocessor for use in mobile products, especially smartphones such as the iPhone or Samsung Galaxy, and tablet computers like the iPad. It powers Qualcomm's Snapdragon, Apple's series of Ax application processors, Mediateks's chipsets, and also the high-volume low-cost feature phones.

The ARM Connected Community, as the ecosystem of partners is now known, has over 1000 companies participating. These partners add value to the ARM architecture and are a formidable barrier to entry into ARM's processor licensing business.

When ARM was founded back in 1990 it had just one licensee, VLSI Technology, who had shipped a total of 130,000 cores. Today ARM has more than 280 licensees who have collectively shipped over 30 billion cores to date. There are 22 million ARM cores entering the market every day. With valuation of under a billion dollars at its IPO, ARM's market cap today is over $22 billion. It currently employs nearly 3,000 people, compared to just 13 employees the day it was spun out of Acorn.

In Their Own Words: Imagination Technologies

Imagination Technologies has a long history as an IP provider and is known for their graphics processors. In fact, if you have a smart phone, it probably has an Imagination processor. In this section, Imagination Technologies tells their story.

Starting in the mid-1990s, graphics technology entered a period of explosive innovation and growth. The world saw the first commercial graphics processors capable of 3D rendering, video acceleration and GUI acceleration, new applications programming interfaces (APIs) for 2D and 3D graphics, and a number of exciting new companies entering the market with innovations that would ultimately move graphics beyond PCs and game consoles into mobile devices.

Hossein Yassaie, CEO, Imagination Technologies, surrounded by some of the many products that use the company's technology.

One of the exciting newcomers to the semiconductor market was a small company called VideoLogic, founded in the UK in 1985. The company originally focused on graphics, sound processing, home audio systems, video capture and video-conferencing systems, using a combination of technologies developed in-house and leading third-party solutions.

Original VideoLogic logo.

VideoLogic's major innovation was in development of a tile based deferred rendering technology (TBDR) for graphics, which it introduced in the mid-1990s. VideoLogic's PowerVR architecture was the first consumer deferred renderer. The basic idea behind deferred rendering is that visible pixels are drawn, and the covered/occluded pixels are discarded. This was a very different method compared to the traditional process at the time, which drew every pixel, even if the rendered output would never be visible. With TBDR, PowerVR processors were able to make better use of system memory, and dramatically increase efficiency.

Business Growth

In July 1994, the company was listed on the London Stock Exchange, first under the name VideoLogic, then later as Imagination Group plc. From that point, the business began to grow rapidly, based on a number of strategic relationships and investments.

The company formed a strategic relationship with NEC in 1995. With NEC, it designed a series of the world's first PC 3D graphics processors based on PowerVR, and the VideoLogic Systems division created branded PC boards using those chips. PowerVR Series1 products, the PCX1 and PCX2, introduced in 1996 and 1997 respectively, were available as the OEM graphics on some Compaq PC models, and as PCI cards from vendors such as Matrox.

PowerVR Series2, also developed with NEC, was integrated in Sega's Dreamcast console, which was released in Japan in November 1998, as well as in Sega's Naomi arcade system. Naomi games found in arcades at the time included House of the Dead 2 from Sega and Power Stone from Capcom. By 1999, NEC had shipped over one million PowerVR 2DC chips to Sega for use in the Dreamcast and Naomi systems.

Sega Dreamcast console, circa 1998.

There were also PowerVR Series2 products for the PC (Neon 250 graphics accelerator) and arcade (as well as Sega's Naomi and Naomi2, there was the R-Cade Vision 250 for the ArcadePC platform).

A strategic relationship with STMicroelectronics announced in 1999 was instrumental in bringing PowerVR technology into dozens of new products. ST's KYRO, announced a year later, was the first full-featured PC graphics and video accelerator based on Imagination's PowerVR Series3 technology. Using TBDR technology, KYRO and KYRO II chips provided excellent image quality and a complete modern feature set at a reasonable cost, enabling developers to create rich environments at high frame rates.

In 1999, the PowerVR 2D/3D graphics processor design was granted Millennium Products status, announced by Britain's Prime Minister Tony Blair as part of the Design Council's Millennium Products initiative.

A Change in Business Model

By 1999, Imagination was creating a large number of innovative technologies, and decided to make these available to the wider market. Under the leadership of CEO Hossein Yassaie, the company formally refocused on intellectual property licensing and changed its name to Imagination Technologies to better reflect the company's activities.

Consequently, the company was split into two operating business units. The PowerVR Technology division developed and marketed PowerVR graphics/video technology, and the VideoLogic Systems division produced a range of innovative and award winning products in the areas of 2D/3D graphics and sound acceleration, home audio systems, electronic music, DVD, digital entertainment, video-capture, and video-conferencing. VideoLogic's consumer product brand soon became Pure Digital, and after that was simply called Pure. Over time, Pure has become a world-leading consumer electronics manufacturer, leading the way in mainstream wireless music and radio systems and entertainment cloud services as well as innovating in new areas such as TV set-top boxes with advanced graphical UIs.

Imagination was awarded the title of 1999 Company of the Year in the prestigious PLC Awards, sponsored by PricewaterhouseCoopers in association with The London Stock Exchange and the Financial Times. The award recognized Imagination for its strong management and long-term strategy. A short time later, in April 2000, Imagination was awarded two Queen's Awards for Enterprise. The Awards for Enterprise Innovation and International Trade were presented to the company's PowerVR Technologies division.

Enabling the Mobile Graphics Revolution

Around that time, the company announced another long ranging strategic decision: it would take its PowerVR architecture into mobile devices.

While even the best mobile computing devices of the late 1990s had little graphical capability, the company was convinced that its technology, which had been designed for low power, could enable a revolution in mobile visual applications.

With innovative differentiators such as TBDR, as well as a low memory bandwidth and low-power advantages, Imagination's PowerVR GPUs were well positioned to lead the mobile graphics revolution. A number of strategic partnerships beginning in early 2001, as well as a new product family, PowerVR MBX, which also launched that year, set the stage.

PowerVR MBX was a complete 2D/3D graphics solution for wireless multimedia devices, with two variants – MBX, which was optimized for speed, and MBX Lite, which was optimized for low power consumption. PowerVR MBX was Imagination's first PowerVR core for mobile devices that included support for the company's proprietary PVRTC texture compression technology. PVRTC significantly decreased the memory footprint associated with texture mapping in GPUs.

Initial MBX mobile licensees included Hitachi, Renesas, and TI, and MBX was a key component of the STMicroelectronics Pocket Multimedia (PMM) platform. Numerous other leading semiconductor companies soon followed, with the platform being licensed by seven of the top ten semiconductor manufacturers at the time.

In 2002, Imagination created Imagination Technologies KK in Tokyo to enable it to exploit further opportunities with Japan's consumer electronics and semiconductor companies. Other major licensees joined with Imagination around this time to proliferate Imagination's technologies into new areas. New broad-ranging license agreements were announced in 2002 with companies including Intel and Frontier Silicon.

Key strategic partners supporting the proliferation of PowerVR included HI Corp., Connect Technologies, and ARM. Imagination also joined the Khronos Group as a promoter member in 2002 to drive development of open standard APIs which allow manufacturers to leverage new graphics capabilities, such as those found in the PowerVR MBX core.

A Broader IP Portfolio

While PowerVR was driving the creation of entirely new categories of mobile products, Imagination's CEO Yassaie was already thinking beyond graphics to providing larger system solutions. In 2000, Imagination acquired Ensigma, a fourteen-year-old private company specializing in Digital Signal Processing (DSP). With Ensigma, Imagination gained expertise and state-of-the-art algorithms in the key areas of audio and speech processing for wireless and Internet communication.

In 2001, Imagination further extended DSP technologies with the launch of Metagence Technologies ('Metagence' was later shortened to 'Meta') and also purchased Cross Products Limited, a company that designed and produced CodeScape development tools for processors.

The Metagence processor architecture leveraged multi-threading to run several real-time tasks on a single processor, rather than using inefficient multi-DSP solutions. The first processor based on the Metagence architecture was the META-1 core. It was integrated into Frontier Silicon's Chorus FS1010, a single-chip Digital Audio Broadcasting (DAB)/audio processor that also incorporated receiver technologies developed by the Ensigma team. The first product to use the Chorus FS1010 chip was Pure's highly-popular sub-£99 EVOKE-1 radio in 2002.

Pure's highly-popular sub-£99 EVOKE-1 radio, circa 2002.

The same chip was used in Pure's products up until about 2005, and was also used in hundreds of other digital radio products from other manufacturers. Later Pure products used follow-on versions of the Chorus SoCs, with newer versions of Meta and Ensigma technologies.

Over time, Meta continued to evolve, getting a floating point unit, higher clock speeds, Linux and Android support and more. As of 2013, Meta had evolved into a leading audio platform that is embedded in numerous generations of products. Meta is also used in many of Imagination's IP platforms for video and communications. The Ensigma technologies have also continued to evolve, and have gone on to ship in tens of millions of devices. CodeScape continues as Imagination's comprehensive suite of development tools, which supports the advanced and unique features of Imagination's programmable IP cores.

2005: A Banner Year for PowerVR Graphics and Video

2005 marked a major milestone with the introduction of the PowerVR SGX GPU architecture. The first implementation of PowerVR SGX was Imagination's PowerVR Series5 scalable and fully programmable multi-threaded universal shader graphics core family. The first SGX cores targeted mainstream and high-performance mobile graphics with state-of-the-art support for 2D and 3D and a feature set that exceeded OpenGL ES 2.0 shader and Microsoft Vertex and Pixel Shader Model 3 requirements. Shaders are advanced effects applied to the graphics image which enable more realistic images to be created. Unlike traditional 3D rendering, shaders are programmable, enabling the content developers' creativity to become the defining factor on how a game, UI or application looks.

2005 also marked the introduction of PowerVR video encoder and decoder IP cores. Since then, the company has introduced five generations of PowerVR VPUs (video processing units), comprised of a balance of hard-coded and programmable elements which combine to deliver efficient multi-standard and multi-stream video decoders and encoders. As of 2013, Imagination's PowerVR video IP had shipped over 600 million units.

PowerVR Graphics Leadership

PowerVR graphics continued to proliferate, and the industry took notice. In 2006, both Intel and Apple invested in Imagination, and the companies have continued to be significant stakeholders in the company. By the end of 2006, there were more than thirty handsets in production from a range of vendors using PowerVR GPUs, including handsets from NEC, Nokia, NTT Docomo, Panasonic, Samsung, Sharp, and Sony Ericsson.

PowerVR's progress and innovation continued unabated. In 2007, Imagination demonstrated the first OpenGL ES 2.0 silicon. By 2008, PowerVR graphics were the de facto standard in mobile graphics, reaching a milestone of having been shipped in over 100 million consumer products. The 200 million unit milestone was reached in 2009. A quarter of a billion PowerVR-enabled devices shipped as of 2010. As of 2013, PowerVR graphics had shipped in over 1 billion devices, making it the most successful graphics technology for mobile and embedded applications.

In 2012, Imagination introduced its latest generation of PowerVR graphics processors. The PowerVR Series6 'Rogue' GPU architecture was built on the maturity and success of the previous five generations of PowerVR graphics IP cores. The PowerVR Series6 GPUs are based on a scalable number of compute clusters, arrays of programmable computing elements designed to offer high performance and efficiency while minimizing power and bandwidth requirements, with an architecture approximately 5x more efficient than previous generations.

PowerVR GPUs are also capable of doing more than just graphics. By supporting compute-based APIs such as OpenCL, Renderscript and Filterscript, the PowerVR architecture delivers vast parallel processing power, increasingly referred to as 'GPU compute.' Using this technology, GPUs will increasingly come to dominate 'heavy lifting' processor-intensive computing as part of heterogeneous SoCs.

With this in mind, in 2012, Imagination joined the Heterogeneous System Architecture (HSA) Foundation as a founder member, together with

AMD, ARM, MediaTek, Qualcomm, Texas Instruments, and Samsung. The HSA Foundation is a non-profit consortium focused on defining and providing an open, standards-based approach to heterogeneous computing.

PowerVR Developers

Imagination's public Graphics SDK (Software Development Kit), first introduced in 2001, played a key role in PowerVR adoption. Designed to enable all software developers to produce games, applications and utilities optimized for PowerVR, the SDK enabled developers to learn by example how to get the best from PowerVR. The SDK has been updated and innovated over time to enable developers to take full advantage of the growing capabilities of PowerVR.

With the launch of the PowerVR Insider program in 2005 and the addition of a comprehensive PowerVR Insider online resource for developers in 2006, the Imagination developer program continued to grow. Today the PowerVR Insider SDK is a cross-platform toolkit designed to support all aspects of 3D graphics application development, specializing in support for devices that contain PowerVR GPUs and enabling users to get the most out of the graphics acceleration hardware available to them. In 2013, the PowerVR Insider community had more than 40,000 members.

A Growing Portfolio of SoC IP

Over time, Imagination has continued to set pace in a range of technologies, bringing to market new microprocessor, DSP, communications and video technologies, with a focus on high performance, power efficiency and multi-standard capability across its range of IP offerings.

In 2010, Imagination announced that it was bringing to market its Flow portfolio of enabling technologies for cloud connectivity. The Flow technology had already experienced success in powering the Pure division's market-leading Flow range of connected audio products.

Today, Imagination's FlowCloud technology includes highly-integrated licensable hardware based on Imagination's market-leading silicon IP and supporting software solutions, complemented by a range of internet-based technologies and a portfolio of cloud-based resources and services together with access to an extensive and growing ecosystem of partners' services and content.

In addition to adding capabilities to its existing IP portfolio, the company also looked for continued areas of expansion. In 2010, Imagination acquired two new companies. First was HelloSoft, one of the world's leading providers of Video and Voice over Internet Protocol (V.VoIP) and wireless LAN technologies. This acquisition addressed the key requirement for network operators in the 4G age to ensure devices can access all different networks, with varied connecting technologies.

Imagination's second 2010 acquisition was Caustic Graphics, a developer of hardware/software real-time ray-tracing graphics technology. Ray tracing is a technique for rendering cinema quality 3D at a level of near photographic realism that is impractical with traditional 3D graphics techniques. As of 2013, Imagination offers this technology in Caustic Professional ray tracing PC boards for content creation professionals, with plans to provide the technology in IP form in the future.

The company's technology portfolio continued to expand with the 2012 acquisition of Nethra Imaging, a semiconductor and systems company focused on delivering video and imaging solutions.

With the addition of these technologies, Imagination continued its focus on building a total solutions portfolio for future SoC designs.

Popular MIPS Architecture Comes to Imagination

In 2013, Imagination completed its acquisition of MIPS Technologies. With MIPS, Imagination added to its IP portfolio one of the most prolific, longest-living processor architectures, greatly enhancing the company's CPU offerings and roadmap.

Over more than three decades, MIPS has powered products including game systems from Nintendo and Sony; DVRs from Dish Network, EchoStar, and TiVo; set-top boxes from Cisco and Motorola; DTVs from Samsung and LG; routers from Cisco, NetGear, and Linksys; automobiles from Toyota, Volvo, Lexus, and Cadillac; printers from HP, Brother, and Ricoh; digital cameras from Canon, Samsung, FujiFilm, Sony, Kodak, Nikon, Pentax, and Olympus; and countless others. MIPS licensees have shipped more than 3.5 billion units since 2000.

At the heart of MIPS is a pure RISC (reduced instruction set computing) instruction set, a clean and elegant solution that leads to lower power consumption and smaller silicon area than other CPUs. MIPS processors feature advanced technologies such as hardware multi-threading, compatible 32-bit and 64-bit instruction set architectures (ISAs), and ISA consistency from entry-level to high-end.

Continuing Innovation

CEO Hossein Yassaie was awarded a knighthood in the 2013 New Year Honors. The award was given in recognition of his services to technology and innovation.

In 2013, the company began preparing for the next stage of its growth, focused on total SoC solutions. With PowerVR graphics and video, this includes support for 4K ultra-HD video technologies, driving GPU compute applications, and enabling the next generation of 3D graphics technology with ray tracing. It also includes enabling low-power, multi-standard connectivity with Ensigma radio processors (RPUs), providing the industry's highest quality of service for V.VoIP and VoLTE through its HelloSoft IP and leveraging the company's FlowCloud technology to enable seamless delivery of services and content between service providers and users through the cloud. Another key strategic initiative is in driving MIPS CPUs to become a leading force in the market.

Innovation also continues in the Pure consumer electronics (CE) division. Building on the strong foundation of its success in radio, Pure has been driving product and platform developments that significantly broaden its market reach to include wireless streaming and internet connected audio, broadcast radio, in-car radio and audio, cloud-based services and connected set-top boxes. Pure is also key to Imagination's partnerships in entertainment and content. In 2012, Pure engaged with Onkyo, VW group, Universal Music Group, Alpine and Pioneer, helping to consolidate Imagination as a significant voice in entertainment technologies.

Imagination leveraged its audio expertise in the development and 2013 launch of Caskeid, a technology that delivers exceptionally accurate synchronized wireless multiroom connected audio streaming. Pure's Jongo system is the first Caskeid-enabled multiroom system to deliver the sync performance and quality of a wired system in a wireless setting. Caskeid works seamlessly with Imagination's FlowAudio cloud-based music and radio service which delivers access to over 22 million music tracks as well as hundreds of thousands of radio stations, on-demand programs and podcasts.

Imagination House at the company's headquarters in King's Langley, Hertfordshire, UK.

The "Market Share Analysis: Semiconductor Design Intellectual Property, Worldwide, 2012" report from market research firm Gartner showed that the third-party semiconductor design IP market grew by 11.2% in 2012, and in that same period, Imagination grew by 36.4%. MIPS Technologies, recently acquired by Imagination, also outpaced industry growth in 2012, growing by more than 17%. For the sixth year in a row, Imagination maintained its position in the survey as the third largest design IP provider, with its overall share growing each year. With MIPS in the fourth position, the companies together comprised 11.3% of the design IP market share.

As of June 2013, Imagination IP has cumulatively shipped in over 5 billion devices, with many of those devices containing more than one of Imagination's technologies.

Chapter 8: What's Next for the Semiconductor Industry?

We've talked a lot about the history of the semiconductor industry, from its nascent beginning with the invention of the transistor and integrated circuit, through the changing business models and technological innovations that shaped the world of electronics we have today. But where are we heading?

Currently, smart phones and tablets powered by highly-integrated SoCs are the largest market driver for semiconductor technology. Even so, over the past 5 years the semiconductor industry has seen relatively flat revenue growth. The following passages are from industry luminaries sharing their vision of what will take the semiconductor industry to the next level of innovation and financial success.

Moshe Gavrielov
President and CEO of Xilinx, Inc.

There is a relentless drive to ever smarter systems consuming ever more bandwidth from communications and computing. There are smarter phones, smarter networks, smarter data centers, smarter factories, smarter cars, and smarter energy, just to name a few. From the consumer to the enterprise, factories and infrastructure, there is more knowledge and use of vision, locations, applications, resources, quality of service, and security. There are ever more analytics required for small or 'Big Data' along with automated control processing, provisioning, configuration, and overall system management.

There is also a trend to ever more programmability in systems, and the devices that support them. There are software and hardware programmable data centers to more holistically manage servers, storage and networking with new open standards, to get more spectral efficiency out of new wireless heterogeneous networks made up of both large and small cells with adaptability to every standard and spectrum, to make factories more adaptable to new models and derivatives, to make vision-rich medical ultrasound systems, driver assistance, and high-definition surveillance systems upgradable with higher resolution cameras and new algorithms for object detection.

Companies that combine smarts with all forms of hardware and software programmability will enable product development teams to maximize their system value and get to market faster with adaptability, reuse, and rapid upgrade cycles. They will enable operators to be more efficient in utilizing bandwidth, maximizing quality of service, and minimizing total cost. They will enable end users to get what they need, when they need it, and wherever they need it.

The programmable logic industry is in the midst of a major shift to provide ever smarter, 'all programmable' solutions to enable these next generation systems. From smarts in the form of IP and embedded processing, to programmability in highly integrated FPGAs, SoCs and 3D ICs, this shift will change the landscape of the semiconductor industry. I guess we are all getting a lot smarter… and all programmable.

Simon Segars
CEO of ARM

At ARM we are dedicated to reducing the barriers to innovation. The growth of the fabless industry has played a crucial part in delivering that vision. It has allowed a rapid evolution of chip designs as we and our partners are able to leverage a significant and shared body of silicon production expertise and knowledge. By centralizing the manufacturing build-out and operations in specialist foundries the industry has enjoyed economies of scale. In turn the foundries can also focus on production innovation enabling the whole industry to benefit in terms of cost and the speed of design iteration. Since we founded ARM more than 20 years ago, over 50 billion ARM-based chips have been shipped by our partners. Without the seamless design-to-end-user supply chain that has evolved through the growth in the fabless industry our vision would remain a dream.

As I look forward today, the fabless industry is even more important as the rate of change we're facing forces each segment of the industry to get even more efficient at what it does. ARM is working alongside the foundries and our design partners to accelerate innovation at every level. Silicon foundries are innovating across the spectrum from new low-power and low-cost technologies at 180 nm to the latest cutting edge FinFET transistors and everywhere in between. On the chip and device side, it's exciting to see how processors are being used in more and more innovative ways with every passing year. The mobile phone industry has spawned an exciting new category of wearable devices. The embedded world is embedding connectivity to create the rapidly growing trend commonly called the Internet of Things. The enterprise world is taking advantage of lower power server solutions. Without the fabless semiconductor industry these leapfrog innovations would not have been possible. Together with the foundry industry, the fabless movement has democratized innovation and created a freedom that has led to the most exciting time the electronics industry has ever seen. New and creative form-factors are being tested—some will thrive, and some will die, as is the nature of evolution. None of this would be possible without the shared and accessible capabilities the foundry industry offers the chip makers and inventors of the world.

It is with thanks to the fabless industry and our foundry partners that we can look to the future, a future in which we will see incredible innovation including products we haven't yet envisaged.

Aart de Geus
Chairman and Co-CEO of Synopsys, Inc.

There are fabulous technology advances underway at just the time when the explosion of applications is gearing up to bring in BIG money into the semiconductor techonomic ecosystem. What we're seeing is the latent pull of the applications market, riding on the back of the past 50 years of semi technology push.

And the waves of applications are just beginning to hit the market shore. No matter what field you look at, you notice a substantial uptick in hunger for silicon: "If you could just make the chip design a little faster/lower power/smaller, we could do it!" In other words, semi innovation will be most visible through utilization. It will be a connected world of "smart everything," where everything, and possibly every one, will have chip-driven computational capabilities and an IP address.

All this "smartness" is already creating massive amounts of data, and the communication and analysis of this "big data" is going to burn through computational capabilities at a rate that we've never seen before. With such a tremendous possibility for expansion, the money from the application world will find its way to support the investments and growth in the semi industry necessary to make it happen.

Our job then, as part of the semi ecosystem is to make it all possible at a reasonable price. This doesn't mean it always has to be cheaper, but for sure we need to enable "better" and "sooner" to keep pace with the speed of innovation we're seeing at all levels of the semi food chain.

The good news is that the outlook for FinFETS on the manufacturing side will gain traction with each FinFET tapeout, and the outlook on the design side is positive as well. From the EDA/IP perspective, we will see great momentum in some near-term roll outs of substantial EDA advances as well as excellent results and progress in systematic IP reuse and assembly to speed up and reduce the cost of design.

Walden Rhines
CEO and Chairman of Mentor Graphics

The history of EDA industry growth has been driven by the emergence of new design challenges. The early generation of schematic capture and simulation was quickly augmented by PCB design, IC place and route and physical verification. In the last ten years, virtually all EDA industry growth has come from totally new design methodology requirements, e.g. sale of IP blocks, resolution enhancement, ESL, formal verification, design for manufacturing and a few more. The best assumption is that the future will evolve as it has in the past, i.e. solutions to new design problems as well as the application of EDA technology to challenges in other areas of design.

With the evolution of IC design into the 14, 10 and 7 nm realm, there will be requirements for analysis of new physical design problems. Examples include reliability, electromigration, thermal effects, stress, EUV resolution enhancement and yield analysis. Even larger will be the adoption of electronic design automation by system design companies that have been able to get by with semi-manual methods in the past. Automotive and aerospace applications are the most obvious since the electronic complexity of cars and aircraft is increasing so rapidly, probably 5% per year or more. How long will it be before we simulate the electronic behavior of an entire car or plane? A long time. But the capability to design and optimize the electrical interconnect, verify correct operation of safety, environmental and security features, manage the trade-offs in cost and weight versus performance and provide a complete electronic data base that can be used by automotive engineering, manufacturing and service is already here and will be a big part of EDA industry growth in the next ten years.

Lip-Bu Tan
President and CEO of Cadence Design Systems

The semiconductor revolution continues to shape the world we live in. And today, a number of exciting trends are driving semiconductor development and demand. These include mobile Internet devices, cloud computing, wearable computing, social media, the "Internet of Things," and more. Over 10 billion mobile Internet devices are in use today (2013) and this number is expected to grow to 50 billion devices by 2020.

While these trends are creating a lot of demand, semiconductor companies are also facing some acute challenges. One is time to market – there is constant pressure to get more designs done more quickly, even though design complexity is skyrocketing. Another is the increasing amount of software that must be created, integrated, and verified along with the silicon.

Silicon technology is increasingly challenging as well. Advanced technology nodes at 20 nm and below require special techniques (such as double patterning), a new type of transistor (FinFETs), and tools and methodologies that can design and verify hundreds of millions of transistors. ICs must deliver high performance, consume little power, and fit into increasingly smaller form factors.

Given these challenges, the semiconductor industry can reach the next level of innovation and success only through deep collaboration. Many years ago, IC design and manufacturing was relegated to a few large IDMs. Today, thanks to the fabless semiconductor revolution, hundreds of companies and design teams worldwide are designing semiconductors. Fabless companies are part of a larger ecosystem that includes EDA vendors, semiconductor IP providers, and foundries. No one in this ecosystem can succeed alone, but we can all succeed together.

Dr. Ajoy Bose
Chairman, President, and CEO of Atrenta Inc.

It is no secret that the semiconductor industry is in transition. Accelerating disaggregation - of manufacturing and content (IP) - leaves less and less opportunity for traditional fabless semiconductor suppliers to differentiate and control. Market growth in China is tantalizing but opportunity for suppliers outside China is at significant risk thanks to a rapidly rising Chinese semiconductor industry, viable at gross profit margins (GPM) of 20%, where Western companies are accustomed to 40%. The Internet of Things (IoT) is another large growth opportunity but will permit only razor-thin margins. Like it or not, the market seems set to be dominated by low margin demand and increasing competition for the foreseeable future, except for those elite few who are able to move up the food chain to offer more total ownership of systems and services.

To survive and prosper in this climate, established suppliers must focus on price-competiveness and differentiation beyond mainstream content. Net R&D efficiency must be enhanced - delivering winners ahead of schedule, cutting losing investments as soon as possible - and that means better controlling and de-risking design. Social policy aside, margin pressure will continue to drive design-offshoring, increasingly to new low-cost regions, demanding even higher levels of design control. Design on older process nodes is another opportunity to reduce cost, if higher levels of integration can be managed through 2.5D and 3D technologies.

There are multiple opportunities for differentiation in content. Reliability and security are likely to be long-term concerns. Ultra-low power will be essential in IoT devices.

Differentiation in both these areas could be a significant barrier to competition, especially from China. Outside mainstream process development, there is likely to be significant demand for advanced sensors, MEMS and even more sophisticated non-traditional technologies, such as lab-on-a-chip. The future is bright for nimble innovators in process and content, but the clock is running out for any suppliers unwilling to adapt.

Jack Harding
President and CEO of eSilicon

One of the few technical tenants of our industry that has crossed over into mainstream vernacular is Moore's Law – the profound observation that chip complexity doubles every 18 months, with a decreasing cost per gate at each process node. For decades this notion has guided an entire industry. The cost per gate now appears to be increasing, with a slowing adoption of new technology nodes. Is this the end of Moore's Law? Will the semiconductor technology roadmap literally run out of road?

The semiconductor industry is renowned for some of the world's boldest innovations. It has solved countless problems with uncanny predictability and this will not stop. In fact, I expect the industry to accelerate rapidly in the coming years thanks to three primary forces.

First, semiconductor technology will become heterogeneous. Thanks to 2.5D and 3D packaging technology it will no longer be necessary to put all system functions on one piece of silicon. This innovation will allow a much wider class of systems to be designed and built in a cost-effective manner.

Second, semiconductor technology will become ubiquitous. Sensor technology and the Internet of Things revolution are examples of this trend. When will all roof tiles be solar? When will all walls contain pressure and temperature sensors, as well as sound transducers? These are obvious requirements for a truly "smart building". A trip to Home Depot the day after tomorrow might be a very different kind of experience.

And third, semiconductor technology will become accessible. The Internet has made a vast array of technologies and information readily available on your mobile device. The semiconductor supply chain is on its way to your hand-held device right now, with the ability to answer highly complex technical questions about chip design not far behind. I'm personally looking forward to the day after tomorrow.

Kathryn Kranen
CEO of Jasper Design Automation

Electronic systems have revolutionized our modern lifestyles, giving us near-real-time access to the collective intelligence of experts on every topic, and the ability to connect and share with virtually anyone, anywhere. Facebook, Amazon, and Google user experiences rely upon connectivity, collaboration, and concurrency, enabled by today's semiconductor technology. We can already foresee self-driving cars and delivery drones; refrigerators and pantries that automatically shop for groceries; and wearable medical systems that monitor and communicate our health and dispense medicines to improve it.

The pace of electronic innovations and the financial success of the semiconductor industry are gated by the scalability and cost-effectiveness of the semiconductor development process. The good news is, electronic design automation (EDA) solutions have improved engineering productivity by 100,000X over the past three decades. Despite those advances, today's semiconductor development process involves many serialized and isolated steps with long latencies and loop-backs. Conventional semiconductor design methodologies are based on a "design-integrate-simulate-implement-oops!-rework" cycle. To date, the semiconductor/EDA ecosystem has not fully taken advantage of the connected, collaborative world it has helped to create.

By reaping what we have sown, the semiconductor/EDA ecosystem has a big opportunity to upgrade the semiconductor development process. What if all designers knew their cross-dependencies and downstream implications (on functionality, software requirements, energy consumption, performance, physical effects, silicon cost, etc.) in real time? True concurrent engineering would require a combination of three things: existing and new EDA technologies; semiconductor design best practices; and big data analytics running on connected cloud computing.

By bringing the innovations of the Digital Age to semiconductor design, we can dramatically lower costs, improve time-to-market, and leverage the real-time collective intelligence of semiconductor and system design teams.

Hossein Yassaie
CEO of Imagination Technologies

Spurring innovation is all about looking for discontinuities and spotting trends across multiple markets. This involves a deep understanding of what drives markets and how technology must be designed to address that. Mobile is perhaps the best example of this; when the first smartphones were launched, even industry insiders wondered whether there was any future for these devices in the context of portable electronics. Smartphones have since become the main consumer computing device —I'm pleased to say due in part to the powerful but low-power graphics technologies that Imagination pioneered.

The consumer market enables everyone in the supply chain to thrive, and this growth can only be sustained through ensuring the best possible user experiences and original, exciting product offerings that make a real difference to consumers. The entire industry must be increasingly aware of what consumers want and need, and focus their engineering towards fulfilling those needs.

Some design implications of the consumers' desire for smart, connected devices are obvious. For the next wave of high-volume devices such as wearables, a CPU with low power credentials is critical and connectivity is highly desirable.

To meet cost constraints, we will also see complete hardware integration —beyond the CPU and GPU. For years, incorporating Wi-Fi and Bluetooth components on the SoC has been a hot topic among integration and design engineers. Now, by incorporating an RPU (radio processing unit) into the SoC, the next generation of devices will deliver a combination of world TV, world radio and complete connectivity.

We understand the value of investing in technology for consumers and society. Our philosophy is to create IP that enables better communication between social groups and delivers the solutions companies need to build better platforms for a range of applications from entertainment to e-health and beyond, using energy and resource wisely.

Fabless: The Transformation of the Semiconductor Industry

Sanjiv Kaul
CEO of Calypto Design Systems

It is humbling to be asked to share your thoughts on the innovation and profitability prospects for perhaps the most dynamic and innovative industry in the history of business. The semiconductor industry has transformed our lives in so many ways that we can't imagine a world without it.

The drivers of growth for the semiconductor industry in the short to medium term are becoming clear. There is a virtuous cycle developing of ever more pervasive mobile internet devices, applications that leverage those devices, and cloud computing to store and analyze all the data collected. As more devices get connected to the internet, we find ways to use those devices to create more value. This is all leading to the "Internet of Things" where everything will be connected to the internet. This market dynamic promises to generate good growth for the semiconductor industry, which stands ready to deliver. The innovations needed to get to sub-10 nm technology keep coming, so there will be no lack of ability to meet the increased demand for silicon. The challenge will be an economic one.

The challenge in the semiconductor industry has been one of profits. Given how capital intensive the industry is, can a company generate enough profits to remain a leader or even relevant? This is one reason there has been so much consolidation in the industry. In the 1980s it looked like no one could stop the ascendancy of the Japanese semiconductor companies. Yet, as the PC became the primary driver, profits and market power shifted to Intel with its X86 architectural lock. Because few other companies could afford the state-of-the-art fabs, we saw the emergence of companies like TSMC and the fabless semiconductor companies. Building on this trend, the internet revolution shifted power and profits to fabless companies like Qualcomm and Broadcom. The interesting question today is, which companies will come out the winners in this round of battles for supremacy? It is hard to predict the winners because the battle is still on, but what happens will have profound impacts on the structure of the industry.

Srinath Anantharaman
Founder and CEO of Cliosoft

The fabless semiconductor business model has indeed helped reduce the cost of getting ICs to market over the past 40 years. The dramatic reduction in capital costs has led to a robust and competitive industry with many innovative startups. To meet the aggressive price and performance demands of the market, design teams continue to grow, tapping into the best talent available across the globe. IP and design service vendors have stepped in to help accelerate development further, increasing the need for effective collaboration and design data management.

The growth of IP and IP reuse will create even greater cost pressure for chip designers, so all the basics for cost-effective collaboration will need to be in place for verifying IP, ensuring quality, and handling the data explosion that new technologies, markets, and products will bring. This data explosion is exacerbated by the task of managing larger, diverse and geographically dispersed teams working closely, securely, and reliably on the same designs. As well as using innovative tools for improvements in power, performance, simulation, etc., design companies will need to ensure that they aren't overwhelmed by the problems inherent in managing volumes of disparate data in real time. All tools and systems along the design chain will need to be tightly integrated to enable seamless collaboration and content control.

The increase in A/MS functionality on ICs has created a broad spectrum of data that needs to be shared across multiple sites and platforms. Real-time communication and collaboration between engineers—at the same site or around the world—reduces the need for expensive, time-consuming meetings, error-prone email communication, and ad hoc data sharing methods. All team members will increasingly need to track the progress of the design, merge their files together as a working set, and easily identify and roll back changes that may not be compliant. A full audit trail, automated with error-proof bookkeeping and gate keeping capabilities, will have to exist to relieve managers and engineers from communication and data management tasks that will become increasingly onerous.

Smarter smart phones, wearable electronics, driver-less cars—the future of electronics depends on EDA tool vendors. We will all be focused on keeping innovation up and costs down as electronics become even more ubiquitous. EDA tools have progressively increased designer productivity to keep up with Moore's law but inefficiencies creep in as teams grow. The next low hanging fruit is to improve team productivity through effective collaboration.

Charlie Janac
President and CEO of Arteris

The semiconductor industry continues on its historical growth path while facing unprecedented near-term change. Our industry is accelerating the trend of bringing computing technology physically closer and closer to the people using it. We started with the mainframe in big data centers, then the minicomputer on the same floor as its users, to engineering workstations and personal computers that sat on our desks, and now to the smart phone that sits in our pockets.

The computing power of these machines in relation to each other has not changed all that much. Arguably the mainframe of the 1970s, the Minicomputer of the 1980s, the Workstation of the 1990s, the Personal Computer of the 2000s and the Smartphone of the 2010s have roughly the same computing power, driven by increasing semiconductor integration and smaller device geometries. The smartphone of the near future will have the same computing power and more functionality than the personal computer of just a year or two ago. Smartphone wireless connections to keyboards and various size displays will mitigate human form-factor issues allowing smartphones to largely do the job of PCs.

So will this evolution stop with the growth of the smartphone? Absolutely not.

As an interconnect IP provider for systems-on-chip, we are seeing major investments in wearable computing where the computers are incorporated in watches, glasses, wrist bands, and clothing. Many previously unconnected objects such as weight scales, thermostats, and tennis racquets will be wirelessly connected to PCs, smart phones, or directly to the cloud. Everything is becoming connected to everything.

This long-term trend will offer both opportunities and dangers—there will be winners and losers. For every Google, Amazon, and Facebook, there will be Walden Books, industry print media, and minicomputer manufacturers. One key enabler allowing semiconductors to scale in size and functionality will continue to be technical innovations in the on-chip interconnect, facilitating the integration of ever increasing numbers of IP blocks and hardware functions.

Where do semiconductors go after the coming wearable computing revolution plays out? This may seem far-fetched but I would not be surprised to see a generation of implantable computing where low power computing, combined with neuroscience and medical device technologies, will make us into even more capable human beings. It will likely happen, but it is hard to predict when. But when it does, semiconductor technology will enable it.

David Halliday
CEO of Silvaco

While many will argue that the "Internet of Things" and "Smart Everything" will be major drivers for Semiconductors, I disagree. While they will be a good side act, both are incremental growth on the existing technology. The Internet of Things, especially, will have low requirements and will mainly be mains powered. It will also suffer from razor thin margins and low costs.

To find the next true driver we need to look much deeper. We need to look to not to ourselves, but to younger generations, as the young have long been the driving consumer. Today they are a connected flock, requiring a constant connection to bombard them with bite size snippets of information from; text, WeChat, Snapchat, Facebook, Twitter, Reddit. For more in-depth content, our young have all but forsaken print and moved to video. This alone guarantees that the mobile revolution is far from over. Looking deeper into the future, Google is already working on autonomous systems as a way to provide a vehicle so that people may remain connected even while traveling. Further, if autonomous cars become practical, why not other autonomous systems? Autonomous systems require staggering amounts of semiconductors; from sensors to the power train and it all needs to be reliable or redundant.

Semiconductor growth has long been driven by the consumer and it will take a new and major consumer revolution like autonomous systems to bring forward the next golden age of semiconductors, other applications will be nice filler but they will not be the main act.

We do face some challenges including the staggering cost of 450 mm wafers and the continual delay in EUV. It is therefore likely that 28 nm will remain the cheapest per transistor node for some time. This will open new areas of opportunity such as 3D and stacked dies. In the analog segment shared MEMS and analog die becoming prevalent and this segment will continue to outgrow digital for some time.

Dr. John Tanner
Founder and CEO of Tanner EDA

I'm amazed and inspired by the rate and pace of technological innovation that has been achieved in our industry since I founded Tanner Research in 1988. The breadth and depth of talented designers and engineers that have delivered breakthroughs to our ecosystem have had a profound impact on business and society; something I'm proud we've been a part of.

I believe connections and collaboration are essential to the future success of our industry. Connections are the necessary but not sufficient component to the success model; they are the seeds of collaboration. Inter-industry, intra-industry and cross-industry connections are required. As the technological and business issues that we face will grow more complex and multidimensional, we'll need to draw cross-domain expertise and capability into the fold. Wide-reaching cross-domain connections will be essential to overcome new challenges and limitations in capability and capacity. Changes to the design workflow will also be an impetus for new connections. We already see that design and layout are no longer able to operate as loosely-coupled entities; they need to connect more directly. This necessitates a richer, more effective interaction between the groups.

Properly cultivated connections are essential for collaboration. While effective collaboration models are in place today (Tanner, like many others, collaborates with foundries for PDKs and with technology partners for extended capabilities), I foresee collaboration models developing across new dimensions of our industry. Perhaps crowdsourcing or collaborative investment models can be applied to help fund new EDA companies; fueling innovative start-ups. A grand challenge issued by The National Academy of Engineering is directed at engineering the tools of scientific discovery; citing the need for engineers and scientists to collaborate in understanding the many unanswered questions of nature. Perhaps this type of collaboration model can be directed towards our industry as a way of popularizing some of our manifold issues and bringing new talent into our ecosystem.

Xerxes Wania
President and CEO of Sidense

Where is the semiconductor industry heading? Good question. I can predict this all day long but at the end of the day it will probably be no more than, exactly that, a prediction. If I could predict the future, I'd be playing the stock market rather than running a semiconductor IP company. I'm sure you have all heard that one before. That being said, I did work with several semi start-ups in the past 25 years and they have all had good exits. I guess I've been doing something right.

I think that the future of the semiconductor industry will be full of surprises. A lot of consolidation will occur as smaller semiconductor and IP companies, lacking marketing and sales resources, will be bought by larger, usually inefficient, companies that are slow in introducing new products.

What I'm really excited about is the electronic do-it-yourself trend— what I like to call the eDIY revolution. If you scan the web you'll find that playing around with inexpensive microprocessor boards like Raspberry Pi and Arduino are becoming very popular, not only amongst the tech savvy but even amongst the high school students who may feel challenged to build their own media centres on a budget. These inexpensive computers are powerful and are becoming easier to program and control. Add to these boards some memory and specialized circuitry for control and automation, and the sky is the limit. I see a significant rise of popularity of these types of boards.

3D printers could be the next high-volume product that every household would like to have. Imagine replacing a broken shower head, a tail light of your automobile, your knife handle or your TV remote button with something you could print on demand. This won't happen overnight but a sub-$200 3D printer should be possible.

The semiconductor industry will grow steadily as the demand for electronic devices continues to grow for the next 10-25 years. I'm quite positive that the growth will come from the Internet of Things, home automation and monitoring, health diagnostics and care, electric/hybrid cars and drones.

Ghislain Kaiser
Co-founder and CEO of Docea Power

The new era of smart/connected devices, the "Internet of things" and organic electronics (converting plastic or glass into a smart surface) opens new possibilities and applications where electronics and software are the base compounds. More and more electronic devices will be part of our life to make it easier, whether it is true or not, that is what the marketing slogan will tell you.

Those new applications will come with new design challenges. Existing concerns will move to the next level of complexity, which will make them critical as new KPIs. Security and power consumption are two examples that illustrate that point. Security will be a heightened concern and it's not just your PC and your smart phone you will be worried about, but also your tablet, car, and home. Power consumption is tantamount as billions of simple daily life gestures will trigger a cascade of activities on chips over the world as your sensors, mobility devices, or smart objects communicate with servers spread across the globe. All that will contribute to drastically increase our energy bill.

On one hand the divide-and-conquer paradigm will be increasingly seen as the right way to reduce the design complexity. On the other hand, as specific concerns are transversal, new design approaches and tools will be required to guarantee an overall optimization and validation while improving gains in productivity. System tools will really emerge and address specific interactions between hardware (chip, board, multi-boards), software, mechanical issues and user experience. That way, the required cost and expertise to design a full application will get lower, which will open opportunities for new start-ups that will only outsource the manufacturing.

Once almost everything and everybody is connected to each other in the future, a new form of collective intelligence will appear. Some might call it Skynet but that is another story about ethic and technology.

Jodi Shelton
Co-founder & president of the Global Semiconductor Alliance

I co-founded the Fabless Semiconductor Association (FSA) 20 years ago with the belief that the fabless business model would define the industry. FSA was the organization that best understood the fabless/foundry transition that led to the acceptance, adoption and finally the domination of the fabless business model, and we remained its advocate for nearly two decades.

In recent years, we were also the organization who recognized that fabless stopped being the differentiator and that companies were implementing pure and hybrid models, and as such, we embraced business model diversity. To keep pace with the industry's evolving nature, we transitioned from FSA into the Global Semiconductor Alliance (GSA).

The semiconductor industry is a main driver of modern technical innovation. With the support of our members, we strive to showcase how the industry offers real and measurable value to society by improving its quality of life and productivity, which impacts and transforms sectors such as energy, medical, communication, education, and entertainment.

Looking forward, there are opportunities that will arise from the industry's pending challenges and I believe this will help determine what is next for our industry. Our industry's top CEOs are concerned with the blurring of the traditional demarcations between supplier, partner, customer and chip designers as well as further consolidation of not only our industry, but also our partners and customers.

It will be interesting to see how the ecosystem responds and changes to an even bigger shift from hardware to software. Many executives today believe that in addition to hardware, students need to heavily focus on software if they want to succeed in this dynamic industry.

And of course, we will see the proliferation of the Internet of Things and even faster data exchange through the cloud. Innovation is key to improving society and our industry is key to innovation. It's an exciting time and I look forward to witnessing how the semiconductor industry will shape the future over the next 20 years.

Gideon Wertheizer
CEO of CEVA, Inc

Low power consumption is set to become the dominant theme across the semiconductor industry in the coming years. The past 4 to 5 years have seen the relentless demand for more processing power lead to higher power consumption and heat dissipation in mobile devices and network equipment. From smartphones and tablets all the way though to wireless infrastructure and data centers, we see a clear requirement for the semiconductor industry to tackle the power-hungry chips and focus on reducing the power requirements and heat dissipation. Moreover, advanced technologies like LTE-Advanced, 4K HD video and the plethora of new, 'always-on' use cases in mobile devices that utilize the microphones and cameras become prohibitive in this environment.

The semiconductor industry now turns to the challenge of reducing power consumption, and one obvious way is to use advanced, low power process geometries with lower leakage current and lower voltage operation. This comes in conjunction with better design practices for memory, standard cells, and mixed signal technologies. Another useful approach relates the chip microarchitecture and, more precisely, using the right blend of the three anchor processors—the CPU, GPU and the DSP—in a highly-efficient heterogeneous system architecture. Today, the emphasis on quad-, hexa- and octa- CPU core processors as the method to deliver more processing power does not address the power consumption issue. A better distribution of the workload in an SoC to correctly utilize a blend of these processors can result in significant improvements in performance and power savings of the overall SoC for a wide range of end applications. For example, utilizing a DSP as the main processing engine for many real-time applications such as face detection, object recognition and noise reduction in smartphones, tablets and smart TVs can realize power savings up to 20X.

This thinking has recently led to the formation of the HSA (Heterogeneous System Architecture) Foundation whose goal it is to improve the ability to program heterogeneous parallel devices and help the semiconductor industry better address the underlying power consumption issues facing the industry.

Grant Pierce
CEO of Sonics, Inc

The semiconductor industry is driven to innovate in face of demand from consumers to integrate a growing number of sensors in ever more complex systems and smaller form factors. This is not news, it is the law of the jungle, it's Moore's law, its More than Moore. And it is even more difficult to predict now what will happen as device sizes approach the atomic level. But, we can predict that innovation will continue. Innovation will answer the call from the market—the semiconductor industry is THAT good.

The Internet of Things (IoT) is today's call from the market that is driving innovation. IoT is causing a dramatic shift in the products that are ultimately purchased by consumers and is a technology game-changer in these important ways:

- IoT brings an explosion in the number of mobile devices, stimulating even more focus on reducing power consumption in all system modes. This problem will be addressed in nearly every level of design, and in IP.
- Many, perhaps most, IoT devices will need to harvest their own power. This will drive innovation in tools supporting concurrent design of electrical and mechanical systems. This will also increase the need for improved mixed-signal design.
- IoT devices will have a multitude of types of sensors, further increasing the need for concurrent electrical/mechanical tool systems and improved mixed-signal design.
- IoT devices will have increasing pressure to incorporate higher security standards. IoT devices will not only control access to physical objects, but to our bodies as well through remote monitors and medical devices. Security will be paramount.
- Time-to-market pressures coupled with the diversity of types of devices will drive the capabilities hardware/software co-design by three orders of magnitude over the next 10 years.

IoT is earthshaking. It will be fascinating to see where IoT takes us, and to be part of changing humanity as IoT is absorbed and utilized. I believe the IP industry will make many pivotal contributions to the fabless semiconductor industry in this area. I can't wait to see this all come to life.

Trent McConaghy
Co-founder and CTO of Solido Design

In John Von Neumann's time, transistors were still a laboratory curiosity. Computation was done with vacuum tubes. Vacuum tubes were horribly unreliable, breaking due to trapped gases, slow leaks, stress on the filament, and more. They'd fail all the time. Teams of techs were employed full-time to replace dead tubes. After trying out vacuum tubes, computer maker Konrad Zuse found them so unreliable that he went back to using electromechanical relays. John Von Neumann saw another way: use just the right math to build reliable machines out of unreliable components. He wrote a paper about this in 1956. But just a few years before, Bardeen, Brattain, and Shockley invented the first practical transistor. Among other advantages, transistors were far more reliable than vacuum tubes. Vacuum tubes got left behind, and with them, most major worries about reliability.

We've now been using transistors for more than six decades. Transistors keep getting smaller, driven by the aims of faster speed and lower power—and higher profit. But our nanoscale transistors have a problem: the size of the transistor is starting to approach the size of atoms. Gates may be just a few atoms thick. With just a few atoms out of place due to manufacturing variations, our beloved transistors become as unreliable as vacuum tubes. This is happening. Yields have dropped and performance gains are stalling. Similar to the teams of vacuum tube-fixing techs, there are armies of testers to calibrate devices after silicon. The economic rationale for smaller transistors is evaporating.

What are we to do if we want to keep driving Moore's Law forward? How about the following: go back to Von Neumann. Listen to the math. Design reliable chips out of unreliable transistors.

This math is there. It's statistics. It's machine learning, which can be viewed as a modern branch of statistics. It's Monte Carlo algorithms and related variance-reduction techniques. It's other statistical-computational tools like moment matching and density estimation. And if you're a design engineer, it's the CAD software that uses these modern statistical tools. The tools do probabilistic analysis to handle variability at the device and circuit level, so that the higher circuit levels can operate as if there isn't an issue. In fact, that's exactly how today's leading semis are already designing. It's essential in order to ship the chips at the bleeding edge.

As we go forward, I anticipate statistics will play an even larger role for semiconductors, in the form of probabilistic cores and more. Many compute-intensive applications, from web image search to speech recognition, use statistical machine learning algorithms. Yet these algorithms almost always run on binary Turing machines, which squander power and speed by having precision when it isn't required. The idea is simple: pull the probabilities up from the device and circuit level, all the way up to the instruction level. The result is probabilistic cores, which operate on probabilistic instructions. These cores can run modern statistical-computational algorithms far more efficiently. Probabilistic cores are complementary to CPUs in the same way that GPUs are. Probabilistic cores are already happening, but they will become far more prominent as ever-smaller devices get more unreliable and the need for more efficient speech recognition and the like grows. Besides probabilistic cores, I expect to see the manufacturing back-end get a whole lot more probabilistic too, especially if technologies like self-assembly see the light of day.

Reliable design from unreliable components has been an aim from Von Neumann's vacuum tube days to our nanoscale devices today. We could try to patch the problem with armies of techs. Or we could listen to the math, embrace the unreliability, and reap the rewards of yield, performance, and speed.

Mike Jamiolkowski

CEO of Coventor, Inc.

The use of automated modeling and simulation technologies has dramatically improved the cost, productivity, and speed with which advanced ICs can be designed. But similar levels of sophistication and innovation have not been applied at the same level to the downstream manufacturing steps, notably process development. This is especially problematic given the complexity of modern 3D IC processes driving Moore's Law forward.

Today, a large portion of the billions of dollars spent on new process technology development is spent on iterative trial-and-error cycles of learning, using many wafer-based experiments in the fab. I estimate the iterative cost of these cycles of learning at $50-$100 million every 2-3 months.

What's required is more predictive modeling of advanced manufacturing processes, much the way IC design architectures and layouts can be analyzed "virtually." By predictively modeling the full fabrication process as applied to any design, a "virtual fabrication" approach would allow the development of 3D models of the resulting structures on wafer, saving millions in unnecessary iterative cycles. These predictive 3D models would hold most of the critical data that is measured during the fabrication process (critical dimensions, defect mechanism, sensitivities, etc.) and therefore eliminate the need for many of the costly experimental cycles.

Virtual fabrication supports the trend of the large fabless IC vendors bringing more Silicon technology expertise in-house to ensure better design technology trade-offs with their foundry suppliers. I believe this type of predictive modeling of IC technology will be the key to sustained progress in semiconductor process development, enabling new levels of innovation similar to how EDA tools and methodologies paved the way for faster and better IC design.

Joe Sawicki
Vice President and General Manager at Mentor Graphics

If you look at what is magical about the mobile revolution, it is driven by a combination of three factors: ubiquitous availability enabled by wireless; the web providing unlimited, user-fed information; and the smartphone providing dedicated computing. The three reinforced each other to produce the biggest explosion in semiconductor demands since the PC.

You don't need to be particularly skilled in prognosis to see the Internet of Things as the next logical extension of the trend. It leverages the three elements that drove the mobile revolution, but adds an entirely new level of information sources allowing us to interact with and pull data from the things around us.

By allowing us to get and control the state of virtually anything, we'll be able to revolutionize how we manage and interact with the world. The home, the factory, transportation, energy, food—all will be deeply impacted by this new idea. It is quite possible that this will drive an increase in productivity that will dwarf that gained by PCs and mobile.

Semiconductor systems enabling this trend will need to respond to difficult cost, size and energy constraints to drive real ubiquity. We'll need to find a 3D packaging implementation that is an order of magnitude cheaper than current interposers. Implementation of these will require an even deeper understanding and ability to model system effects, putting a premium on tools that enable design and verification at the system level and engineers that can use them. Cost constraints will also drive innovation in test to ensure that multi-die package test doesn't explode part cost.

Subi Kengeri
Vice President, Advanced Technology Architecture,
GLOBALFOUNDRIES

In the coming years, our industry will be driven by a proliferation of new applications and services such as social networking and cloud computing, Big Data analytics, wearables, and the Internet of Things (IoT)—a cluster of nodes connected wirelessly to enable a smarter world. Between 2010 and 2020, world population will grow modestly from about 7 billion to 8 billion, while during this timeframe the number of connected devices is expected to grow from about 13 billion to 50 billion, an annual growth rate of 15%. The growth of these new applications will require innovative solutions based on semiconductor technology. For example, IoT presents some extremely challenging design and manufacturing issues. An IoT chip in the past was brought to market in about 3-4 years, but future requirements will be closer to 1-2 years with simultaneous reduction in cost to below one dollar per chip. The signal chain is just as complicated as a cell phone, but at about 1% of the cost and 1/1000th the size and power consumption. New innovations across the semiconductor ecosystem will start to emerge to address such challenges, including ultra-low power operation and energy harvesting to enable remote operation. Together, these technologies will permit devices to be powered indefinitely from sustainable sources, opening the door to ubiquitous sensing, environmental monitoring, and medical applications.

Along with these new applications and technologies will come increased complexity, demanding an even greater level of collaboration across the semiconductor ecosystem. The cost of design and manufacturing at the leading edge will continue to skyrocket, driving an increased transition to the fabless/foundry business model. But gone are the days when a fabless company can develop a design in isolation and simply hand it over to a foundry partner. The challenges of the coming years will require true "collaborative device manufacturing" partnerships to deliver continued innovation. The era of Foundry 2.0 has arrived, and only companies that are willing to collaborate early, deeply, and openly will survive and thrive in this environment.

Martin Lund
Senior VP of the IP Group at Cadence Design Systems

The increase in design complexity and the pace of innovation staggers us today. But imagine reading this passage in just 3-5 years. We'll look back on 2014 and think our pace was comparatively leisurely! How do we get there from here?

Consider: in the cloud, 3.3 zettabytes of traffic increase double digit percent annually; hundreds of millions of mobile devices enter the market each year; and Internet of Things devices are expected to top 50 billion by 2020.

Demand coupled with time to market, design complexity, and standards and technology evolution are flipping tried-and-true design techniques on their heads. At 90 nm, you could expect about 18 IP blocks per design. At 14 nm? Try 123 IP blocks. Hardware must be optimized for software at the earliest stages of the design process; design teams needs to embrace a holistic view of the end systems across all functional disciplines to get to market in time.

IP selection must be reconsidered in the coming years using methodologies that enable increased levels of automation for SoC customization. It needs to be reconsidered because designers are dreaming big. They're dreaming big because Moore's Law's effect on functional integration means successful electronics companies scamper up the food chain to deliver more value to their customers and capture greater margins.

Once a cog in a big churning wheel, IP vendors today define complete subsystems and make big contributions to the full-chip requirements because of all the hardware and software functionality they deliver. Their customers, in turn, are defining system requirements for their customers.

Integration, design automation and hardware verification—embracing a holistic system-design view—are the keys to improving engineering productivity as we race into the future.

Rich Goldman
Vice President of Corporate Marketing and Strategic Alliances at Synopsys

During the past 50 years, the semiconductor industry has driven the fastest pace of change in the history of humankind. We commonly know this as Moore's Law. Today, the ICs kids wear on their feet to control their light-up sneakers contain more computing power than existed in the entire world in the year of my birth, 1960. We can barely fathom the changes that will occur during the next 50 years, but we can be assured the semiconductor industry will drive them. After all, what other industry has improved by 100 billion times in 50 years?

The demise of Moore's Law has been predicted many, many times, and so far everyone who has bet against Moore's Law has lost. I'm not going to do that, but eventually the run of silicon must come to an end, and certainly far before 50 more years pass. Probably in the next 20. Why? Keeping up with the pace of Moore's Law continues to get more and more difficult, both physically and economically. The economic challenges are forcing us to advance in three axes by developing new methodologies and new devices, and by experimenting with new materials. As we always have, engineers will find the way forward by combining advances in all three areas, including die stacking (3D IC) and FinFETs.

Engineers will also need to make radical changes in materials as we move to new nanotechnologies, or they will need to start basing computing on photonics or quantum effects. Moore's Law ruled 50 years ago. Moore's Law rules today. And one way or another Moore's Law will continue to rule as the semiconductor industry drives the pace of advancement. It's the fastest pace in the history of humankind, and it will continue to result in capabilities that we can barely imagine today, even in science fiction.

Raymond Leung
VP of SRAM at Synopsys

The semiconductor industry is maturing and will face a lot of challenges moving forward. In the last several decades, we saw unanimous adoption of new process technologies and the only variable was the adoption timing. The adoption curve followed the bell shape but now, we are beginning to see 'late' adopters not really moving forward at all, or at extremely slow pace. Initially, it looked like the bell shape curve becoming two smaller bell shapes and now the adoption looks more scattered. So, my assessment is that instead of being a technology driven industry, it is more an application driven one now. The challenges are to meet the application needs; may it be costs, power, specialties (like high-voltage or embedded flash), form factor, or performance.

Richard Goering
Veteran EDA editor and author of the Cadence Industry Insights Blog

Much has been written about what's required for continuing success in the semiconductor industry. Deep and early collaboration among participants in the IC design supply chain—including IP providers, EDA vendors, and foundries—is a must. EDA tools need to raise the level of abstraction for design and verification, and enable design reuse. IP providers must offer progressively more complex hardware/software "subsystems" that can be quickly verified and assembled.

What may be lost in this discussion is the importance of imagination and innovation. We need new ideas and new technologies. FinFETs are a good example—here's an idea that came from academia, and it is becoming a reality, with the promise of huge performance and power benefits below 20 nm. We need more "next big thing" ideas, especially as we start to consider what will be needed below 10 nm. What's the role of optical electronics, quantum computing, and carbon nanotubes? What will make EUV cost effective?

The most important industry collaboration is one that will attract a new generation of designers who are interested and excited about solving such challenges. As we grapple with the problems of today, let's lay the groundwork for the bright ideas of tomorrow.

Don Dingee
Semiwiki.com blogger

Most eyes in the semiconductor industry today are focused on mobile SoC development and the path to 14 nm. Just over the horizon lies even bigger opportunity. In the Perfect Semi Storm, four forces collide: the lightning of the Internet of Things, the thunder of programmable logic, the waves of MEMS sensors, and the winds behind the "maker" movement.

The Internet of Things lights up with mobile platforms as the user on-ramp, connected by social experience spanning the planet. Demand for MCUs powering new connected devices explodes, from around 25 billion annually to hundreds of billions each year. Data centers in the cloud expand in new directions, featuring vastly more efficient servers with workload-tuned processors at lower power.

Programmable logic rumbles beyond only the huge FPGA into more SoCs and MCUs, allowing fine-grained customization of I/O and signal processing in every device. Applications like embedded vision, packet processing, and encryption are no longer out of reach for low-power platforms. More uses develop as more designers embrace customizable hardware, and IP for smaller devices grows.

MEMS sensors roll into everything, everywhere, with sensor fusion algorithms powering context. New fabs and processes create smaller, smarter, multifunction sensors with integrated MCUs and wireless radios. Advanced energy harvesting, particularly piezoelectric ceramics and photovoltaic cells, and sophisticated thin film rechargeable batteries help power the tide.

Makers set sail with this inexpensive, incredibly functional hardware and a treasure trove of open source software to new lands of discovery. The beat shifts from PC era multi-year release cycles to a constant swirl of innovation paced by application markets and crowdfunding. Startups form and rule the seas, operating under their own flag and plundering the slow. The Perfect Semi Storm is coming. Prepare to be blown away.

Luke Miller
Semiwiki.com blogger

Given the emergence of stacked silicon interconnect (SSI) and the like, we can expect fully integrated analogue and digital solutions in one package. I foresee internal banks of ADCs feeding programmable logic that can feed banks of DACs. RF/Digital ICs will be real, as well as the integration of wireless, once again leveraging SSI.

We will bid farewell to DDR/LVDS as device pin-outs cannot keep up with the insatiable data demands. We will welcome the Gigabit Serial age with the market dominated by HMC and JESD204 and other competing solutions.

The far future of connectivity will be the replacement of copper with optical IO at the device level. Board traces will be fiber and the gigabit serial challenge of long data transmission at high rates will be overcome. GT speeds will be well over 100 gb/s per lane—which will just be the start—and this will enable the move into terabit signalling that would be impractical over copper.

As voltages drop and densities over 80 billion+ transistors emerge by 2020, EMI, SEU, SER and other bit phenomenon will be at the forefront, as well as design verification and design automation in the EDA realm. As we step out further, electron/quantum computing will unleash a technological world that will blow our minds, solving mysteries and problems once thought impossible.

Erik Esteve
Semiwiki.com blogger

The development of the semiconductor industry has been amazing and we see semiconductors being used almost everywhere. The main driver today is clearly the mobile industry with the fastest consumer adoption ever seen for any products, from television to cars. There is no doubt that semiconductor pervasion will go on and increase in many industry segments, and that new applications will appear, like IoT or wearable communicating products, but…

Historically the price per gate decreased about 35% per technology node. But in the very latest technology it is flat or increasing slightly. Meanwhile the development costs associated with new chips are increasing exponentially from one node to the next.

Will we see the emergence of more fabless start-ups in the future, or concentration of the semiconductor industry through acquisitions? Will we see vertical integration, already effective at Apple or Samsung, going down to EDA or IP companies?

An obvious example of a consolidated industry is automotive. Will the semiconductor industry behave like the automotive industry did when consolidating? Today, application processors for mobile are developed in a very dynamic and competitive industry, and can be compared with a classic high-quality car like the 1958 Cadillac Eldorado. Do we want a consolidated semiconductor industry with less competition that might lead to chips comparable to a 1984 Cadillac Cimarron?

Innovation has always been the driver for a highly dynamic, creative and successful semiconductor industry. I hope that industry actors will be wise, so we leave room for start-ups to emerge and innovation to flow. Such creativity will be the enabler of the post-smartphone products of 2030 or later, keeping the semiconductor industry as dynamic and successful as it has been up to now.

Daniel Payne
Semiwiki.com blogger

The semiconductor industry has several parts that combine to form an ecosystem: equipment makers, fabs, SoC designers, semiconductor IP, EDA software, and embedded software. Most of my experience is in the fab, SoC design, EDA software and IP space, so I'll limit myself to those areas.

We'll continue to have a handful of large foundries and IDMs that can afford the multi-billion dollar costs of process development, and the marketing focus to keep the fabs filled and running at a profit. Companies like Samsung and IBM have figured out how to be both an IDM and foundry, so there's some hope that even Intel can join their ranks by adding a selective foundry approach instead of a broad foundry business model. To manage new process development costs I expect that the collaborations between companies will continue.

Intel is leading by getting FinFET (aka Trigate) into production first, with the foundries playing catch-up on this technology that will gradually supplant bulk CMOS. A second process technology gaining traction is FD-SOI, which is an attractive alternative to FinFET especially where lower-power is a requirement.

For 30+ years we've seen in EDA that three large companies dominate by providing 85% or so of all software revenue, and that trend will continue. Smaller EDA companies will focus on emerging new segments that are small enough to not attract the big three, and yet large enough for them to grow and become profitable. An EDA startup can expect an exit strategy of being acquired by the big three or merging with other small EDA companies in order to broaden their software portfolio. The

EDA software business has been acting like a mature industry with single digit growth for years now, so the prospect of an EDA IPO seems dim.

Most of the new growth has been in semiconductor IP, which has grown beyond just standard cells, memory and processors to include AMS and sub-systems with software and drivers. When EDAC reports revenue numbers for the EDA industry it also includes semiconductor IP, which is where the revenue growth continues to be seen. I'm expecting more acquisitions in this space as companies like Synopsys and Cadence are keen to grow through adding strategic IP providers.

My hope is to see some new, disruptive nano technology emerge that will disrupt the status quo and incremental improvements. In my lifetime I've witnessed the transitions from discrete Bipolar to integrated NMOS to CMOS to FinFET, watching transistor counts explode from a handful to billions on a single chip. I wonder if I'll see the first SoC with a trillion transistors on it?

Pawan Kumar Fangaria
Semiwiki.com blogger

In the current era, the internet has been the key catalyst that led to mobile devices and smartphones, which have been the leading drivers of semiconductor business. The next revolution in the semiconductor industry will be driven by the Internet of Things. We will see a rise in the level of automation, security, privacy, convenience, and other things we can't imagine now. The whole world will be connected through multiple links.

This connectivity will bring further transformation to the semiconductor industry. Moore's law will take on another dimension; more like "number of transistors per unit volume will double and price will halve every two years." Ultra-nano technology nodes will have to deliver products faster, with lower cost, higher performance, and lower power. 3D-ICs will become ubiquitous. This level of scaling, faster at lower cost, will have to be supported by Virtual Fabrication, which will be the new catalyst of semiconductor development. Actual foundries will fabricate only near-to-perfect designs, and provide high yield at minimum cost.

This transformation will be supported by MEMS, which will take the lead role in the coming era of Internet-of-Things. Most of the semiconductor designs will have MEMS sensors embedded into them. The design automation tools will also transform in order to handle the design and verification of complete systems that include MEMS and ICs with a high level of prediction about yield, reliability, power, thermal tolerance etc. Low power use will be critical and will demand newer hardware and software solutions. Renewable energy sources will gain momentum for securing battery life.

Considering this massive evolution, data storage requirements will explode, giving a further push to cloud computing. Newer forms of high-density memories and intelligent ultra-high capacity servers will evolve.

Paul McLellan
Semiwiki.com blogger and author

It is clear that Moore's Law will continue, in the sense that we know pretty much how to build chips all the way down to 7 nm. And by the way, 7 nm doesn't have anything on it as small as 7 nm, it's just a name. But what is a lot less clear is whether the cost per transistor will continue to drop or whether 28 nm will be the cheapest process not just to date but forever. As for memories, at some point the capacitor that holds the data in a DRAM will also stop scaling. 3D chips might turn out to be another game-changer but currently they are also too expensive.

We are entering an era where the economics that have driven semiconductor for five decades may be coming to an end. Even if each process is still a little cheaper per transistor, which Intel reckons is the case for them, we are not going to have the sort of 10,000X reduction in price that resulted in our smart-phones having more graphics capability than million dollar flight simulators of 30 years ago. They will never become like credit-card calculators, so cheap they can be used as giveaways.

This is something that the general public and the press have not yet noticed. Everyone just considers that Moore's Law will continue since the iPhone 5 is more powerful than the iPhone 4. But that is a very limited market that isn't affected by a few dollars extra for an SoC. The iPhone 5 is not cheaper than the iPhone 4. The PC you used to buy for thousands of dollars is now just a few hundred. But that trend is not happening with iPhones.

On the other hand, something may change. Some technical breakthrough in semiconductor technology or EDA may break the logjam. Carbon nanotubes? Directed self-assembly? Optical interconnect? Quantum computing? EUV may even work. Or maybe this time it really is different.

Daniel Nenni
Semiwiki.com blogger and author

As I see it, the biggest challenge for the semiconductor industry moving forward is economic. I have no doubt that the fabless semiconductor ecosystem will clear the technical hurdles required to continue down the process node path. I do however think we are underestimating the effect of the increasing financial burden of modern semiconductor design and manufacturing.

The advent of the fabless business model opened the doors to all comers who had an idea and a modicum of capital. The shear competitive nature of this paradigm shift brought us innovative products only Gene Roddenberry could have imagined at palatable costs. Unfortunately, as our industry matures, consolidation comes knocking and that is when the doors of innovation start closing again.

A reduction in the work force comes with consolidation and that is a very difficult trend to reverse. Today less than 5% of the work force is engineering and science based. How do we attract new talent when college graduates see the semiconductor industry as old and boring compared to the likes of Google, Facebook, or Twitter?

Attracting new capital is also a problem. According to the GSA January 2014 Market Watch Report, in 2013 semiconductor funding deals decreased 63% from what was raised in 2012. The total number of fabless semiconductor deals closed during the same period decreased by

30%. And the average value per deal decreased 13%. If you really want to know what will finally kill Moore's law the answer is economics, absolutely.

We as an industry do not do a great job of communicating our amazing story to the outside world. That is the real motivation behind SemiWiki. com and this book; to remind everybody that the fabless semiconductor ecosystem made semiconductors what they are today—a critical part of modern life.

Printed in Dunstable, United Kingdom